The University of Minnesota Press

gratefully acknowledges the publication assistance

provided by Mr. and Mrs. Richard G. Gray, Sr.

D0222916

Handbook
of Animal
Radio-Tracking

L. David Mech

U.S. Fish and Wildlife Service
Patuxent Wildlife Research Center
and
Departments of
Entomology, Fisheries, & Wildlife
Ecology & Behavioral Biology
University of Minnesota

University of Minnesota Press • Minneapolis

Published by the University of Minnesota Press,
2037 University Avenue Southeast, Minneapolis, MN 55414
Printed in the United States of America.

Library of Congress Cataloging in Publication Data

Mech, L. David.
    Handbook of animal radio-tracking.

    Bibliography: p.
    1. Animal radio tracking. I. Title.
QL60.4.M43 1983   599'.0028   83-6733
ISBN 0-8166-1222-6
ISBN 0-8166-1221-8 (pbk.)

This book is dedicated to
William W. Cochran, *Illinois Natural History Survey,*
who, many of us believe, has done more
to further the field of animal radio-tracking
than any other single person.

CONTENTS

PREFACE

"Why don't you write a handbook on radio-tracking?" asked Rajan Mathur, a field officer with India's Project Tiger. That was in February 1980 during my second trip to India with Dr. Ulysses Seal to teach Project Tiger biologists techniques for immobilizing and radio-tracking animals. The trips were sponsored by the World Wildlife Fund, Project Tiger-India, and the Office of International Affairs of the U.S. Fish and Wildlife Service.

The suggestion made sense. Several times I had had occasion to teach other biologists what I had learned through some 20 years of experience in radio-tracking a number of species in a variety of areas. Some information could be found scattered in manufacturers' catalogs and in technical journal articles. Much of the information that a beginning radio-tracker needs to know, however, seems never to have been put down on paper. It just gets passed around by word of mouth or learned anew by each biologist.

To remedy this situation, I have attempted to synthesize the information available that would be useful to researchers interested in beginning a radio-tracking project, particularly those from developing countries. Presumably, biologists elsewhere who have never used the technique would also find it useful. If this book is studied carefully and its advice is followed in setting up a radio-tracking project, the researcher should have a good start on an exciting, productive, and rewarding venture.

It is my hope and intention that this book, even

though I may have left a few things out through over-sight, will minimize the need for researchers to consult other people who have used the technique before they can begin new projects. I know that that is what Rajan wanted, though I doubt that he realized it would take so long to get it.

(Readers may address correspondence to me at the following address: U.S. Fish and Wildlife Service, North Central Forest Experiment Station, 1992 Folwell Avenue, St. Paul, MN 55108.)

ACKNOWLEDGMENTS

Like any collection of information, this book is based on the work of many people. In particular, it depends on the drive, motivation, and ingenuity of a number of pioneers who were able to envision the revolutionary effect that radio-tracking would have on wildlife research. These were all busy people who were willing to gamble their time and resources. They could have safely and productively spent their time using traditional techniques to conduct their research, but instead they bet on a long shot that yielded great returns.

Bill Marshall of the University of Minnesota, for example, attached a $5000 ($10,000 in today's funds) transmitter to the back of an old domestic chicken that he stuck out in the woods. When that worked, Marshall confidently launched the transmitter on the back of a ruffed grouse; he never heard from it again. Undaunted, Bill tried again with a second transmitter and a second grouse. That bird flew smack into a tree and broke its neck. Marshall finally triumphed over the technological problems of this new field, however, after teaming up with amateur radio technician Sid Markusen, and they quickly moved far beyond tracking chickens.

At about the same time, the Craighead brothers, Frank and John, found a willing electronics experimenter in Joel Varney; they began radio-tracking grizzly bears. Meanwhile in Illinois, biologist Rex Lord expressed his wishes to electronics wizard Bill Cochran. Cochran, at that time working on the electronics for some of the army's first satellites, could not believe that biologists were not already regularly radio-tracking animals. He rigged up a transmitter, and he and Lord were in business following cottontail rabbits. These people and others had already developed radio-tracking by the time I came along to help exploit the technique.

As for my own experience with radio-tracking, I want to thank Dwain Warner and John Tester of the Bell Museum of Natural History at the University of Minnesota for giving me the opportunity to learn the technique at the Cedar Creek automatic radio-tracking station. In addition, the following organizations have contributed to my research since then, almost all of which has involved radio-tracking: Macalester College, the U.S. Fish and Wildlife Service, the U.S. Department of Agriculture Forest Service North Central Forest Experiment Station, the Mardag Foundation, the World Wildlife Fund, the Big Game Club, and the Weyerhauser Foundation.

The following electronic specialists over the years have helped me to solve various electronic problems: Bill Cochran, Larry Kuechle, Dick Riechle, and Ralph Schuster.

An early draft of the manuscript was critiqued by Bill Cochran, Larry Kolz, Orrin Rongstad, Jerry Storm, Dick Huemphner, Dave Smith, and Jon Bart, and they offered many valuable suggestions for improving the handbook. Any errors remain my own responsibility, of course.

I also thank Holly Hertel for drawing the illustrations.

Handbook of Animal Radio-Tracking

Uses of Radio-Tracking

Radio-tracking is a revolutionary technique for studying many kinds of free-ranging animals. By March 1979, one of the leading commercial suppliers of radio-tracking equipment had sold over 17,500 radio collars. Numerous species, from crayfish through elephants, have been studied with this technique in widespread regions of the world, from pole to pole and in most major countries of the world.

Even snakes, turtles, lemmings, thrushes (Figure 1), and whales have been forced to divulge their secretive habits and wide-ranging travels to modern-day scientists wielding radio-tracking equipment. Polar bears (*Ursus maritimus*) wander the top of the Earth, while their movements are surveyed via orbiting satellites. In a few areas of the world, subpopulations of several species interact at the same time computers record and plot their minute-to-minute movements. The most recent space-age development in this field is a radio collar that anesthetizes a wild animal on command from a radio signal controlled by a biologist observing the animal at a distance.

What is radio-tracking, and how does it work? How can it help answer questions about specific species being studied, and how can one get started using this technique? This handbook should answer these and many other questions about radio-tracking.

Radio-tracking determines the locations of animals through the use of a radio receiver and directional antenna that trace the source of a signal coming from a radio transmitter attached to the animal. Radio-tracking

Figure 1. Birds as small as thrushes, which weigh about 30 g, have been radio-tracked; usually, the transmitter is glued to the bird's back after some of the feathers have been clipped short (Raim 1978).

offers two main advantages of extreme importance to researchers: (1) it allows precise identification of individual animals, and (2) potentially it allows a researcher to locate each individual as often as desired. Given these two abilities, think for a moment about all the different types of data that could be collected and questions that could be answered by using this technique. No doubt many of the following uses come to mind.

LOCATIONS AND MOVEMENTS

Every time a researcher locates an animal by radio telemetry, he or she acquires a piece of information. The researcher knows that a specific individual is at a particular point at a certain time. Depending on the rate at which locations are detected, several types of information about the individual can be learned. For example, frequent locations throughout the day or night provide excellent insight into an animal's daily move-ment patterns. Even just knowing where an animal started its travels during one part of the day and ended them later is instructive (Mech et al. 1966).

Determining one or two locations per day over a period of weeks gives a good approximation of the extent of an individual's home range. When such data are collected from several members of a population, insight is gained into the spatial ecology of the species. If locations are precise enough and if good habitat information is available, then habitat selection by a species in a given area can be studied (Kohn and Mooty 1971; Nelson 1979).

Longer movements by individuals are especially re-vealing. For example, dispersing animals, although they may have to be tracked by vehicle or even aircraft, provide excellent information about season, direction, distance, and duration of dispersal; age of dispersers; new areas of settlement; and mechanisms of coloniza-tion in new areas (Mech and Frenzel 1971; Van Camp and Gluckie 1979; Fritts and Mech 1981; Berg and Kuehn 1982).

Even longer movements than those shown by dispers-ing animals are involved during migrations of many birds and mammals, and these movements can also be followed via radio-tracking. Many aspects of migration can be studied, such as triggering conditions, migratory path, duration, weather, time of day or night, altitude, and wintering range. Probably the most dramatic exam-ple of the advantages of radio-tracking in studying bird migration was afforded by the tracking of a peregrine falcon (*Falco peregrinus*) for over 2100 mi (3360 km) from Wisconsin in the northern United States into Mexico (Cochran 1975).

Actual experimentation with migration, orientation, navigation, and homing can also benefit considerably from radio-tracking (Evans 1971; Cochran 1972). In such experiments animals can be moved long distances to strange areas and then released, and the manner in which they find their way back home can be determined (Rawson and Hartline 1964; Weise et al. 1975; Walcott et al. 1979; Fritts et al. 1984).

When enough locations of an animal are obtained, and especially if they are either taken at regular intervals or randomly, one can determine the intensity with which the animal is using various parts of its range (Turkowski and Mech 1968). Are certain areas used more than others? If so, what are the features of those areas? In the case of carnivores, such areas may contain concentrations of prey. With herbivores, areas of intense use may indicate better habitat or forage.

BEHAVIOR

Sometimes, from the signal alone, a biologist can infer certain types of activity or behavior from a radio-tagged animal. This is easier with some species, some types of signals, and some types of antennas than with others, and such inferences must always be checked by watching the animal while listening to the signal. Using a long "whip" antenna (p. 20) on ruffed grouse (*Bonasa umbellus*), for example, Marshall and Kupa (1963) were able to determine whether the birds were resting, walking, flying, feeding, or drumming. With most types of radio packages, it is easy to judge whether the animal is active or resting from just the degree of variation in the signal. Additional circuitry and/or sensors can provide further information, such as whether a bird is flying or what its tail posture is (Kenward et al. 1982).

By collecting data at regular intervals throughout the day and night, an investigator can identify the activity patterns of the subject. The degree to which precise patterns can be determined for each species depends on the rate at which the researcher gathers the telemetric data. With an automatic radio-tracking system in which locations were obtained every 45 seconds (Cochran et al. 1965), it was possible to determine exactly when an animal began moving for the day or ended its activities (Mech et al. 1966).

Figure 2. Homing in on a radio-tagged animal such as this leopard and then observing it is one of the many ways that radio-tracking can provide information.

One of the simplest and yet most elegant ways that radio-tracking can greatly assist the behaviorist is merely by allowing him or her to find the subject animal, identify it, and then observe its behavior (Figure 2). In such projects, the telemetric part of the work is critical but takes only a short time. Most of the data are gathered while the subject is being observed (Mech and Korb 1978; Bertram 1980; Rogers and Mech 1981; Ballard 1982).

A special use of telemetry in this way involves studying predation. One can find and identify a specific predator and record from day to day the animal's predatory attempts, at least during daylight (Mech and Frenzel 1971; Mech 1977a). A night-vision scope allows the collection of data at night.

By recording the number of prey taken by an individual radio-marked predator, a researcher can determine approximate consumption rates for that individual (Mech 1977a). Consumption rates for different individuals can then be compared. Furthermore, the relative proportions of different species taken by the predator can be determined; and in some cases, the age, sex, and condition of the prey can be learned (Mech and Frenzel 1971; Kolenosky 1972; Fuller and Keith 1980; Fritts and Mech 1981).

MORTALITY AND SURVIVAL

Studies of predation can also be conducted from the standpoint of the prey. These studies require radio-tagging a large number of prey animals in a given area (Mech 1967; Cook et al. 1967, 1971; Beale and Smith 1973; Dumke and Pils 1973; Brand et al. 1975; Franzmann et al. 1980; Ballard et al. 1981; Nelson and Mech 1981). The age, sex, and condition of each individual is determined at capture. Then, the investigator monitors the locations and activities of the prey; when an animal suddenly stops moving, he or she must then home in on the animal and determine whether it has been killed (Figure 3). The use of special activity-monitoring collars (discussed later) can alert an investigator within hours after an animal has died, and the type of predator involved can usually then be determined.

Using a similar approach, one can study survival rates (Trent and Rongstad 1974; Hoskinson and Mech 1976) and various causes of mortality. Often, by collaboration with a veterinary pathologist, the investigator can determine the precise cause of death as well as the general condition of the animal before its death (Mech et al. 1968; Mech 1977b; Carbyn 1982).

MISCELLANEOUS STUDIES

When researchers have come to understand how telemetry works and what its basic advantages are, they then discover many novel uses that can greatly enhance their studies. An excellent example of the indirect use of radio-tracking is in finding dens and nests. It often is very difficult to locate the dens or nests of various species, and this difficulty has hindered the systematic

Figure 3. By homing on this acrylic radio collar, the author was able to determine that a raptor had killed the rabbit wearing it.

study of such sites. However, radio-tracking a breeding bird or mammal during the denning or nesting season usually leads a researcher to such a site. Observations around these areas also yield valuable information about the productivity of the animal subjects (Mech 1977b).

Another good example is the use of telemetry to study dormancy and hibernation in various species. For example, by locating bears in the northern parts of their range, one can study the times the bears go into winter dormancy and the types of dens they choose (Craighead and Craighead 1972). Then, by examining the animals in their dens during dormancy and just before their emergence in spring, one can also determine their over-winter weight loss. Because bears give birth during winter denning, a check in spring also provides information about cubs.

With some species, census techniques can be greatly enhanced via the use of radio-tracking. When enough

Figure 4. A radio dart enables a researcher to home in on an animal when the creature runs off before the drug takes effect (Lovett and Hill 1977).

deer (*Odocoileus virginianus*) are radio-tagged in an area, for example, their observability from the air can be tested, and the results of an aerial census in the same area can then be adjusted accordingly (Floyd et al. 1979, 1982).

A more recent application of radio-tracking is in tranquilizing animals with darts. Because most anesthetics used in darts require 10 to 20 minutes to immobilize the animal, it is often difficult or impossible to find a darted animal. However, a tiny radio in the anesthetic dart (Figure 4) allows the researcher to home in on the radio signal when the dart is stuck in the animal and to walk right up to the creature (Lovett and Hill 1977). Before using this technique, one must make sure that the darts have barbed tips so they remain in the animal after the drugs have been discharged. This technique also allows one to find darts that missed an animal, an advantage that is becoming very important as the cost of darts increases.

History of Radio-Tracking

It is difficult to say just who started the use of radio-tracking (Kimmich 1980). No doubt the idea occurred to several people at about the same time (p. xi). The first publication on the subject seems to have been by LeMunyan and others (1959), who used a short-range implantable transmitter. Marshall and others (1962) followed them with an article on radio-tracking porcupines (*Erithizon dorsatum*) via external transmitters, and several other workers described their radio-tracking experiences in the 1963 volume edited by L. E. Slater, *Biotelemetry: The Use of Telemetry in Animal Behavior and Physiology in Relation to Ecological Problems.* Nevertheless, the first really workable radio-tracking system, on which current technology and techniques are based, was designed by W. W. Cochran and R. D. Lord, Jr. (1963).

Many improvements in basic radio-tracking hardware and techniques have emerged since 1963 (Cochran 1980; Kuechle 1982). These have included more efficient transmitters (p. 16), more effective materials for encapsulating and protecting transmitters from the weather and the animals (p. 42), and lighter weight batteries (Kuechle 1967). Probably the most significant advance has been the commercial availability of ready-made transmitter collars for a wide variety of species (Appendix III).

STATE OF THE ART

Although radio-tracking was first applied to land mammals and birds, eventually creatures more difficult to work with became subjects of studies using this technique. Marine mammals (Siniff et al. 1975), fishes (Winter et al. 1978), frogs (Jansen 1982), turtles (Carr 1965; Legler 1979; Schubauer 1981), alligators (Smith 1975), and snakes (Osgood 1970; Ikeda and Oshima 1971) have all been radio-tracked.

As the electronic components of transmitters have become increasingly smaller and lighter, more and more studies of smaller species have been carried out. Even ghost crabs (*Ocypode quadrata*) (Wolcott 1980b), land crabs (*Geocarcinus ruricola*) (K. C. Zinell, personal communication, 1982), and crayfish (*Procambarus clarkii*) (Covich, 1977) have been radio-tracked (Figure 5).

Besides improvements in the design of radio trans-

Figure 5. This ghost crab has been capped with a radio transmitter and flashing light-emitting diode for visual location at night. (Photo by T. G. Wolcott.)

mitters, a number of exciting variations on the basic theme have emerged. One type of radio collar uses a special sensor to detect whether an animal is active or not and changes the signal accordingly (Knowlton et al. 1968). A second detects when an animal has ceased moving for a specified period and then alters the signal, thus indicating when the animal has died (Kolz et al. 1973; Kolz 1975). Another type of transmitter indicates mortality by using a thermistor to measure an animal's temperature and then reduces the signal's "beeping," or pulse rate, when the temperature drops sufficiently (Stoddart 1970). A fourth variation on the basic radio transmitter employs a strategically placed sensor to modify a radio signal when the animal urinates (Charles-Dominique 1977).

In a few areas of the world, automatic tracking systems (Figure 6) monitor the movements of many animals simultaneously and a computer plots out their movements on a map (Cochran et al. 1965; Deat et al. 1980). A system has even been developed to compress the movement data and display the movements on a

Figure 6. Automatic radio-tracking systems such as the Cedar Creek system used by the University of Minnesota can monitor up to 52 animals simultaneously and record their locations every 45 sec (Cochran et al. 1965).

movie screen so that several days' movements can be compressed into a few minutes (White 1979). This development allows detection of information in the data that might not show up with a static analysis.

Another development, still incomplete, is the transmission of physiological data such as body temperature, heart rate, and blood pressure from free-ranging animals. Years ago, heart rate (Eliassen 1960) and respiration were monitored in flying birds (Lord et al. 1962), and more recently heart rates have been recorded from free-ranging elk (*Cervus canadensis*) (Ward and Cupal 1979) and bighorn sheep (*Ovis canadensis*) (MacArthur et al. 1979). Nevertheless, many technical problems remain to be solved before physiological telemetry becomes as useful as radio-tracking (p. 80).

The satellite tracking of animals (Warner 1967;

Buechner et al. 1971) also is still in its infancy, with only polar bears (Kolz et al. 1980), sea turtles (*Caretta caretta*) (Stoneburner 1982), and dolphins (Kuechle 1982) being tracked at present (Kuechle et al. 1979). It is conceivable, however, that technological advances in the next decade will make satellite tracking of smaller animals practical.

One of the latest advances in radio-tracking is the capture collar (Mech et al. in press). This is a standard radio-tracking collar with a receiver attached to it that can be controlled by a biologist sending out signals from an airplane. When the correct code is received by the capture collar, one of two darts at the top of the collar fires and injects up to 1.5 cc of drug into the animal. If the first injection fails, the biologist can trigger the second one as a backup. Each time the signal is sent to the collar, the collar's transmitter responds by increasing the pulse rate of the signal temporarily so that the operator can determine that the signal did reach the collar. The collar has been used successfully in the field on wolves (*Canis lupus*) and in captivity on the black bear (*Ursus americanus*), cougar (*Felis concolor*), deer (*Odocoileus virginianus*), and tiger (*Panthera tiger*) (Figure 7).

As early as 1967, free-ranging baboons (*Papio doguera*) were anesthetized remotely via injection capsules implanted subcutaneously and attached electrically to a relay in a backpack operated via radio signal (Van Citters et al. 1967; Van Citters and Franklin 1969). The capture collar recently developed, however, is self-contained, eliminates the need for implants or lead wires, and is commercially available. (See Appendix III.) This makes it applicable to a wide range of species.

Figure 7. This happy ground crew has just recaptured a wolf darted by radio signal from an aircraft. The signal directed the anesthetic capture collar (held by the second person from the left) to dart the wolf (now carried in pack by the third person from the left), and the crew then homed in on the wolf's collar via portable receiver and antenna (held by the first person on the left); in case of partial drugging, a dart rifle is carried as a backup (by the first person on the right).

The Radio-Tracking System

Radio-tracking systems have two major parts: a transmitting system and a receiving system. The transmitting system consists of a transmitter, a power source, and a transmitting antenna, and these are usually attached to an animal in a collar or a harness. The receiving system is made up of a receiver, a power source, a receiving antenna, and an operator or a recorder. A necessary third element of the radio-tracking system is a human interpreter, who may be greatly aided by a computer.

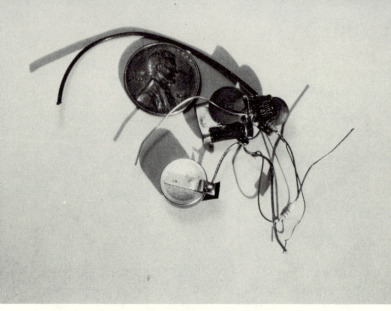

Figure 8. Shown is an example of a commercially available radio transmitter and battery, which are compared with a U.S. penny for size. A transmitter with such a battery would transmit for about 3 weeks.

TRANSMITTING SYSTEM

Transmitters

Commercially available transmitters weigh as little as 1.4 g, and measure only 7 by 10 mm (Figure 8). They have been used on such small animals as mice (Mineau and Madison 1977), lemmings (Banks et al. 1975), and thrushes (Cochran et al. 1967). One component of the transmitter is a quartz crystal, which determines the transmitter's frequency. Usually, each transmitter used in a study occupies a unique frequency, like the different radio stations on commercial radio. Thus, to tune in any specific individual of a group (or population), a researcher merely turns a dial on a radio receiver and hears a different individual on each "station," or frequency.

The frequency of a transmitter refers to the frequency with which the signal oscillates or vibrates during 1 sec. The wave form follows a basic sine curve, with each frequency having a different repetitive cycle time. One full cycle of a sine curve is the portion of the curve from a particular point on the curve to the next identical point on that curve, for example, from peak to peak or from trough to trough (Figure 9). The number of cycles per second is the frequency; for example, 150 megahertz (MHz) indicates a frequency having 150 million ("mega-") cycles per second. To replace the cumbersome phrase "megacycles per second," the term "megahertz" has been adopted; thus, 150 MHz means 150 million cycles per second.

In most telemetric applications, the frequencies of the signals are usually only discriminated to thousandths. Thus, for example, a frequency of 150.346 MHz represents 150,346,000 hertz (Hz), but the zeros are usually dropped. In talking about the difference between the frequency 150.346 and the frequency 150.351 MHz, one can say that there is a difference of 5 kilohertz (KHz), meaning 5000 Hz. Most frequencies used in a study should be kept at least 10 KHz apart for best discrimination on most receivers.

Other electronic components used in the transmitter include a transistor, a capacitor, a resistor, and an

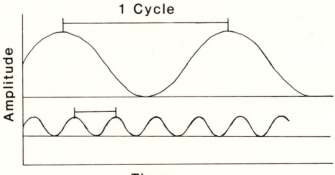

Figure 9. Radio wave forms: the upper graph shows a relatively low frequency, and the lower graph depicts a higher frequency (i.e., one with more cycles per second).

induction coil, each of a specific value. When combined in the correct circuit with a crystal, they constitute an oscillator that when powered produces a signal. Some transmitters include amplification stages that increase the signal. A transmitting antenna launches the signal into space for long-distance propagation. Basic circuits for transmitters have been published by Cochran and Lord (1963), Tester and others (1964), Anderson and De Moor (1971), Corner and Pearson (1972), Kolz and others (1972), Anderka (1980), Kephart (1980), Lotimer (1980), Thomas (1980), Beach and Storeton-West (1982), Kuechle (1982), and others. More efficient circuits have been developed by the manufacturers of telemetric equipment, but these remain trade secrets.

Power Supply

Three types of power supplies are commonly used with radio-tracking transmitters: mercury batteries, lithium batteries, and solar cells. For many years, almost all radio-tracking transmitters used zinc, mercury, or silver cells. Because mercury cells come in the widest variety of sizes and shapes, they are still of value for special uses. However, the lithium cell is only about half the weight of the mercury cell for the same electrical current capacity, measured in number of "ampere hours." Because the ultimate life of a transmitter depends on the capacity of its power supply, the size and capacity of the battery is of critical importance. Furthermore, for most species the size of the transmitter itself is negligible, so it is primarily the power supply that determines the size and weight of the entire radio collar or harness.

The third type of power source is the solar cell, which is a photovoltaic device that produces an electrical current from light (Patton et al. 1973; Williams and Burke 1973). Regular use of solar cells began about 1978, but several suppliers in the United States now produce them (Figure 10). The value of solar cells is obvious: theoretically, they should power the transmitter

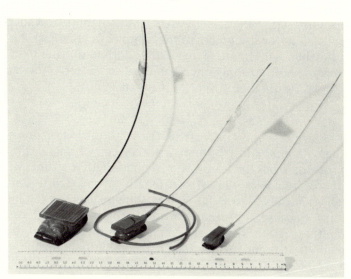

Figure 10. These solar-powered transmitters can be used with birds; similar transmitters mounted on collars are used for mammals. (Photo courtesy of Wildlife Materials, Inc.)

for as long as they remain intact. The main disadvantage is also evident: they will only power the transmitter during the day, and they must be protected from chewing and weathering.

A solar cell can be attached to a rechargeable nickel-cadmium battery, so that the solar cell can store excess energy there during the day. For many applications, this arrangement has proven highly satisfactory, especially with species that spend much of their time in the sun. However, because nickel-cadmium batteries can only be recharged a limited number of times, they have not always proven satisfactory. One other possibility is to attach both a solar cell and a lithium cell to the transmitter. This combination allows the solar cell to power the transmitter during the day, thus saving the power from the lithium cell and then allowing the lithium cell to power the transmitter at night. However, the nighttime transmission life is limited by the capacity of the lithium cell. The effectiveness and reliability of solar cells alone or in combination with electrochemical cells probably will improve markedly during the next few years, thus making their application useful for an increasing number of species (Church 1980).

Antennas

Another important part of the transmitting system is the antenna. Most antennas are pieces of metal that protrude from the transmitter. Their length is a function of the frequency range (wave length) of the transmitter, which ideally would be at least one-quarter of a wave long. Because an antenna of ideal length would be too long to attach to an animal, a compromise must be made that reduces the transmitter's potential range. For a 150-MHz transmitter collar, for example, the compromise antenna is about 29 cm long.

Ideally, the antenna should be straight and point toward the sky from the back of the animal (Figure 11). However, most antennas protruding this way are eventually broken off as the animals plow through the vegetation, or they may be chewed off by associates of the radioed animals. On some species, these so-called "whip" antennas may last for the full life of the collar. On the social carnivores, such as the wolf, however, the antenna must be protected inside the collar. This arrangement compromises the antenna's radiating abilities in two ways. First, it reduces the antenna's effective length because the ideally straight wire must be curved to fit into the collar; and second, it then lies close to the animal's flesh, which decreases the radiated energy and reduces the range of the signal. Nevertheless, most researchers agree that is is better to have a highly reliable antenna with a reduced range than a more efficient antenna that might break off and produce almost no signal after a few months.

A second type of antenna is known as a "tuned loop." This metal antenna is made to fit into a collar and can actually be the collar. In general, large loop

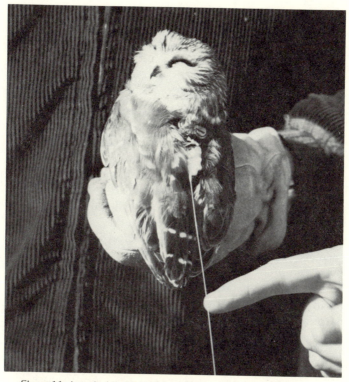

Figure 11. A typical "whip" antenna protrudes from a transmitter on the back of an owl.

antennas are more effective than small loops. However, since the tuned loop is made to a certain size for a specific transmitter, it must not be adjusted significantly in circumference; thus, a collar of a given size must be held until an animal having about that neck size is captured.

RECEIVING SYSTEM

The receiving system consists of (1) a battery-powered radio receiver, (2) a receiving antenna, (3) either a mechanical recorder or a human operator, and, ultimately, (4) a human interpreter. The function of this system is to detect and identify the radio signals from the collared animal and to deduce the location of the animal, as well as other information.

Receiver

Several makes and models of telemetric receivers are commercially available. They differ from the standard broadcast radios primarily in their detection sensitivity. This is necessary because the signals from wildlife transmitters are usually very weak. It is only because of the transmitter's low power requirement that its batteries can function for so long. Assisting the receiver in its capture of low-power telemetric signals is a receiving antenna, which will be discussed later.

Regardless of the make or model of telemetric receivers, they all have several features in common: a power source, an on/off switch, a frequency selector comparable to a station selector on a standard broadcast radio, a signal strength ("gain") control, a jack for earphones, and a jack for an antenna lead. Other features may include a signal-strength meter that moves a needle when a signal is picked up, a battery voltage check, and a jack for a recorder.

The face of one of the most commonly used receivers, the CE-12 (formerly the LA-12) (Figure 12) can be used as an example of the control panels on receivers. Most of the controls are labeled clearly and need no explanation. However, some further explanation about setting precise frequencies is required. This receiver is tuned to a transmitter frequency through a combination of three knobs. The receiver can be purchased with one, two, three, or four frequency bands, each covering a 300-KHz frequency spectrum. If the receiver has four bands (covering a total range of 1200 KHz), it may have a frequency range of, for example, 150.800 to 152.000 MHz (1 MHz = 1000 KHz); then band 1 would cover from 150.800 to 151.100 MHz, band 2 from 151.100 to 151.400 MHz, band 3 from 151.400 to 151.700 MHz, and band 4 from 151.700 to 152.000 MHz.

Figure 12. Shown is a CE-12 (formerly LA-12) radio-tracking receiver. (Photo courtesy of Custom Electronics of Urbana, Ill.)

Within each band there are 12 channels, each covering 25 KHz and selected by a channel-control knob. Each specific frequency within these 25-KHz channels is detected by the fine-tuning knob, which spans the 25-KHz range. Thus, for example, to tune in a frequency of 151.115 MHz, the controls would have to be set as follows: band 2, channel 1, fine tuning about 3. However, the frequencies of some transmitters tend to drift in reaction to temperature and battery changes, and receivers tend to drift in where they pick up a given frequency. Therefore, one might have to tune the fine-tuning knob from about 2 to 4 to be sure of detecting where the frequency comes in.

The sweep control on the LA-12 or CE-12 receiver automatically sweeps across about plus or minus 5 KHz of a selected range in seeking the desired frequency. To use the sweep, the operator turns the band switch, channel selector, and the fine-tuning knob to the right positions for the actual frequency and then flips the sweep switch on. Over about 10 sec, the receiver's frequency tuner will sweep across plus or minus 5 KHz from the point where the fine-tuning knob is set. This sweep option, however, should not be used for detecting weak signals, especially during aircraft searches for lost animals, because the sweep may pass the signal so rapidly that it may not be heard unless it is quite strong.

Most telemetric receivers are able to distinguish signals as close together as 10 KHz. For example, a transmitter with the frequency 151.115 KHz can usually be distinguished from a signal of 151.125 KHz. With such spacing it is possible to handle over 100 frequencies on a four-band CE-12 receiver. However, unless such close spacing is necessary, transmitter frequencies should be kept about 20 KHz apart because of the transmitter and receiver drift problems discussed above.

Receivers other than the CE-12 have different tuning arrangements. On some you can dial in the actual frequency with a set of digital dials. Others use display tubes to indicate the frequency as it is tuned on control knobs. Still others have push buttons, each set to a different frequency.

Most receivers have a signal-strength meter and a small speaker as well as a jack for monaural headphones. The meter and speaker are useful for dealing with strong signals or for testing transmitters in the lab. However, because of ambient noise, it is usually necessary to use earphones when animals are being tracked in the field.

The signal-strength meter is useful in homing from aircraft. Because such tracking involves high speeds, signal strength changes quickly. Thus, it is helpful to be able to assign numerical values to signal intensity and thus to compare signals received over a few seconds as stronger or weaker. The ability to assign a value to the strength of the signal, which is what the signal-strength meter allows, greatly facilitates such comparisons (p. 64).

The gain control can increase or decrease the signal strength or the ability to receive a weak signal. It differs from the volume control that is included on some receivers in that the gain control changes the receiver's meter reading as well as the actual volume of both background noise and signal. The volume control gives no greater sensitivity for picking up signals.

For each person there is a point on the gain control at which he or she can detect the signal. Turning the gain control any higher than that point will make both the signal and the noise seem louder. However, turning up the gain beyond the point at which the signal is detectable will not allow one to detect a signal at any greater distance. This is important to note, because high gain settings on at least some receivers can produce noise levels that damage human hearing. This factor is especially important in an airplane or a vehicle when high gain settings may be necessary to overcome noise and when prolonged monitoring for signals may be necessary. Special earphones for use aboard aircraft are available to help block out extraneous noise; this then allows one to detect a signal with a lower gain control.

Receivers have either of two kinds of antenna jacks to which antenna lead connectors are attached (Figure 13). The smaller, designated a BNC type, has a "bayonet" mount onto which the connector is pushed and then twisted about a half-turn to a lock position. The PL-259, or UHF, type is larger and has threads onto which the connector must be screwed. It is important to know what type of antenna jack a receiver has, because antennas come with either type of connector on the receiver end of their leads. Thus, one must purchase the type that fits the particular receiver; otherwise, an appropriate adapter will be needed to couple one type of connector with the other type of jack.

Receiving Antennas

Almost any piece of metal touched to the antenna jack on a telemetric receiver increases the receiver's ability to pick up a signal. Thus, a long vertical piece of wire can be very useful as an antenna. Such an antenna, if a half of a wave length long and connected at the

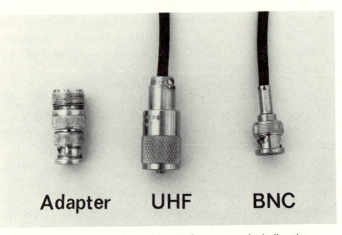

Figure 13. Two commonly used types of connectors including the BNC (right) and UHF or PL-259 (middle); also shown is an adapter used to connect a lead with one type of connector to the other type.

center of its length, is called a dipole. A dipole is less directional and, if properly designed, can be omni-directional. A typical use for an omnidirectional antenna would be for a receiving station placed in the center of a study area to monitor the presence or absence of a signal from the pickup area or to monitor the signal for activity. The main disadvantage of a dipole antenna is that, without moving it from place to place, it is difficult to determine the direction from which a signal is coming. Furthermore, it is not as powerful for its size as other types of antennas.

A second kind of commercially available receiving antenna useful in the field is the tuned loop. This can be a round, oval, or diamond-shaped piece of metal with its size a function of the frequency and with a variable capacitor for proper tuning. Such an antenna is bidirectional, meaning that it can receive a signal best from either of two directions corresponding to its plane. To overcome the problem of ambiguity in direction, one either must walk in one direction or the other

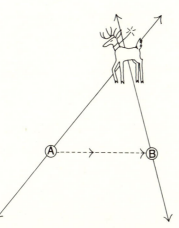

Figure 14. Radio-tracking antennas are often bidirectional; that is, they indicate a signal from two directions, 180° apart. An antenna at position A or B might show a signal traveling from the radioed animal or directly away from it. However, by walking from one position to the other and determining in which of the two directions the bearings cross, the radio-tracker can identify the true direction to the animal.

while listening to the signal to determine whether it gets louder or stronger or must travel more or less perpendicularly to the antenna plane and check the direction again. After a certain distance, which depends on the distance to the signal source, a new attempt at determining the direction of the signal shows the correct direction via a converging pair of bearings (Figure 14).

With many loop antennas the direction of the signal can be determined better through the use of the null of the signal rather than the peak (p. 53). The null is sensed through the center of the loop, perpendicular to the plane of the antenna through which the peak is detected. The directionality is still ambiguous, since the null might be coming from the direction the operator is facing in looking through the null or from the opposite direction. The real advantage in using the null to detect the signal direction is that it is usually sharper than the peak of the signal, so a more precise bearing can be obtained.

A third type of receiving antenna is the Yagi, which

Figure 15. A Yagi antenna receives signals from all sides, but the strongest signal comes from the direction of the radioed animal when the antenna is held with its shortest elements pointed toward the animal.

consists of a boom with several elements attached perpendicularly to it (Figure 15). The length of the elements and the distance between them is a function of the frequency, and each element is progressively shorter along the length of the boom. The longer the boom and the more elements, the greater the sensitivity or pickup range of the antenna and the sharper the directivity. Although larger Yagis at the lower frequency are cumbersome for carrying through the forest or brush, there are commercially available collapsible Yagis that are far more convenient for field use. With a Yagi, the signal is strongest when the antenna is held with the shortest elements toward the transmitter and the longest toward the operator (Figure 15).

Other kinds of receiving antennas are the collapsible H-antenna (Adcock) and an M-Yagi. They give less range than the collapsible Yagi and, like loops, are bidirectional, but they are somewhat more convenient to carry and yield more range than loops.

Accessories to antennas include a lead-in cable connecting the antenna to the receiver and connectors for each end of it. The lead is an electronically shielded

Figure 16. An antenna mounted on a tall mast or tower requires a long lead-in cable to the bottom, where it attaches to a receiver; therefore, a special coaxial cable with lower electronic loss (RG8/U) should be used.

(coaxial) cable designated RG58-U. If the antenna is going to remain mounted outdoors at a considerable distance from the receiver (Figure 16), then a lower-loss cable is desirable, for example, a RG8/U.

The connectors attach the cable to the antenna and to the receiver. The two types in common use have already been described. The important point to remember is that the connectors on the cable must be compatible

Figure 17. (A) The outdoor station automatically monitors and records the presence or absence of a radioed animal in the area. An antenna pointing toward the target area is mounted overhead, with a coaxial lead running down into a weatherproof box. A sunshade is also useful for preventing the box from overheating.

Figure 17. (B) The weatherproof box contains a receiver, a power source, and a recorder.

with the antennas and receivers. When all the equipment is ordered from one company, the connectors are usually compatible. However, when different suppliers are used, the type of connectors needed should be specified. In addition, it is wise to have extra adapters, both male and female types, to convert from one connector type to another.

Recorders

In many field studies, the only recorder is a human being with pen and notebook, and in developing countries this may be the most economical way of recording data. However, with studies of animal activity (p. 72) or with automatic monitoring systems, signal-strength data can be recorded on various kinds of commercial recorders. The usual type of data-collecting system that uses such recorders also includes an antenna, a receiver, a power source (such as an automobile or motorcycle battery), and a recorder (Figure 17). These are usually housed in a weatherproof box with an antenna lead to the box from the bottom (so that rain does not follow the lead into the box).

Strip-chart recorders operable on DC current come with two types of motor drives. One type is less expensive but has relatively poor voltage regulation. This power variation makes the chart paper move through the recorder at a different rate than indicated on the paper, so the net result is that the time indicated on the paper is inaccurate. Furthermore, as each day passes after resetting, the error accumulates. The second type of motor drive costs two to three times as much as the first but is far more accurate. Gilmer and others (1971) discussed the details of hooking up a recorder-and-receiver system.

The best type of battery to power an automatic recording system is the kind that is made for frequent discharging and recharging, such as that often used in powerboats.

Radio Frequency

Radio-tracking equipment, like other types of radio equipment, comes in various frequencies (p. 16). Usually, each radio used in a study varies in frequency somewhat from all the others in that same study. This allows the individual identification necessary to keep track of each animal. Furthermore, there are several major frequency ranges commonly used, and their properties differ. In choosing a frequency range for a study, one should consider the following factors: (1) commercial availability of equipment, (2) legality of operation, (3) transmitting properties, and (4) compatibility or conflict with other studies.

The frequency ranges usually available from most telemetric equipment manufacturers are 32 MHz,

49 MHz, 148 to 152 MHz, 163 to 165 MHz, and 216 to 220 MHz. Although equipment can sometimes be ordered in other frequency ranges, special orders usually take much longer, and there are more complications when nonstandard frequencies are used.

For tracking fishes, much lower frequencies are used. Much fish tracking is actually done via ultrasound methods rather than by radiotelemetric techniques (Stasko 1975; Stasko and Pincock 1977).

The importance of legal considerations in choosing a frequency range varies with the country. Some countries, such as Great Britain, have extremely strict regulations covering telemetric frequency ranges (Skiffins 1982). Many countries, however, have laws regulating or allocating frequency ranges but are extremely lax in enforcing those regulations on scientists using animal radio-tracking. The reason for this is that the low power output of most telemetric transmitters does not interfere with other radio reception. For example, the common type of hand-held walkie-talkie has a power output of 5 w, whereas most telemetric transmitters have a power output of about 1 mw, or one five-thousandth of the power of a walkie-talkie. Thus, the chances of other radios interfering with the telemetric transmitters are much greater than vice versa.

The second reason why many countries do not strictly enforce laws regulating the use of radio frequencies for animal telemetry is that most of the laws and regulations were set for other types of radio use, types that might be in conflict. Because of this, many projects using radio-tracking fail to request official permission for any particular frequency range. Nevertheless, governmental agencies in the United States, for example, regulate frequency use. The responsible radio-tracker would consult other radio-trackers in the study area to determine the best approach to the application and registration of frequencies.

Of course, regardless of whether frequencies are officially approved, it is in the best interest of all researchers to coordinate their frequency choices with those of other radio-tracking projects in the same area. Two main points bear consideration in this respect. First, in countries where radio-tracking is just getting started, it is important for the various projects using the technique to choose the same frequency ranges if their study areas are far enough apart. The reason for this is simple: equipment can be exchanged among the projects.

Generally, telemetric equipment can only be repaired by its manufacturer. This means that, if a receiver, for example, is dropped and damaged, it must be returned to the manufacturer for repair, which could take several months both because of shipping and delays in repair and part replacement. This is good reason for each project to have at least one spare receiver at all times. Nevertheless, even with such a spare, problems can develop requiring the borrowing of equipment from another project. When the frequency ranges are compatible, such borrowing is possible.

The second consideration to be made is that, when radio-tracking projects are located close to one another, there is the possibility of individual researchers picking up signals from animals in another project. This is sometimes advantageous. If an animal strays too far from one study area, a researcher can ask that another project follow it. On the other hand, if communication among projects is poor, a signal from one animal can interfere with the signal from an animal on the same frequency in another project. Confusion or inaccurate results then become real problems. I know of one researcher in Minnesota who thought he was homing in from the air on a radio-collared wolf. The signal, however, turned out to be from a bald eagle (*Haliaeetus leucocophalus*) that had migrated from Missouri!

Obviously, the problem is greatest with studies involving migrating birds or other long-distance travelers. Often, it is quite possible for two or more projects operating close together to use the same general frequency range without interference if the researchers get together and decide which subrange of frequencies each project is going to use. As indicated above, for any given receiver over 100 individual frequencies can be used for the radio transmitters. Since most projects involve fewer than 25 radio-marked subjects on the air simultaneously, this range of frequencies allows considerable latitude for each project. Again, good communication is essential to this process.

Signal propagation characteristics are among the most important considerations in choosing a frequency range. Generally, the lower frequencies tend to work better for transmitting through heavy vegetation or areas of variable terrain. The signals have less tendency to bounce off trees, cliffs, and other natural barriers. However, the Yagi receiving antennas for lower frequencies, such as 32 and 49 MHz, are large and cumbersome. Furthermore, it is very difficult to design efficient transmitter antennas for these frequencies. The intermediate frequency ranges, such as 150 or 164 MHz, tend to bounce much more, but they are more efficient and require smaller Yagi receiving antennas. Further information about the advantages and disadvantages of various frequency ranges is available in Mackay (1970), Cochran (1980), and Sargeant (1980).

Pulse Rates

Although some radio-tracking transmitters use a continuous-wave signal that sounds like a whine, most transmitters made for radio-tracking emit energy in bursts, or pulses. Such transmitters produce a "beep, beep, beep" as heard through the receiver. These pulses can vary in rate and duration (width). The slower the rate of pulsing and the narrower the width, the longer a transmitter can last for a given battery weight. Signals with pulse rates of less than about 60 per minute are difficult to locate. Pulse widths less than about 20 msec are often easy to miss when one is trying to tune in a signal. However, special applications may call for modifying either pulse width or rate to other values, and most telemetric equipment manufacturers offer other various options. One should always record the pulse rate of each transmitter attached to an animal, for the rates usually differ and could be used to determine the identity of a signal in case of overlapping frequencies.

Range

The range of a telemetric transmitter depends on many factors, including the following: (1) its power output, which in turn depends on the weight and type of battery, the voltage, and whether or not there is an amplifier stage to the transmitter; (2) the tuning of the transmitter in relation to the body of the animal; (3) the transmitter antenna, including its length, which is a function of the frequency, its position, and whether or not it is a loop or whip; (4) the height of the animal above ground; (5) the elevation of the animal above the general terrain; (6) the type of vegetation; (7) the soil type; (8) the sensitivity of the receiver; and (9) the type, elevation, and number of elements in the receiving antenna.

Voltage can be important. Some transmitters consist of only a single stage, an oscillator, which may require only 1.35 v. Other transmitters that have an amplifier stage usually require about twice the voltage. Thus, a doubling of the battery weight is necessary to obtain as much life as with a single 1.35-v transmitter. This may be accomplished by either hooking two 1.35-v batteries in series or by selecting a single cell of about 2.7 v. It should be evident that the two-stage transmitter gives a longer range but for a battery of a given weight allows only about half the life as a 1.35-v, single-stage transmitter.

Because transmitters are usually attached close to the body of animals, by harness or collar, the electrical properties of a living animal influence the efficiency and tuning of the transmitter antenna. For peak radiated power, the transmitter and its antenna must be tuned to each other and to the subject animal. If it is tuned before it is attached to the animal, its placement on the animal will tend to detune it and decrease its efficiency. Most commercial suppliers of transmitters understand this fact and tune their transmitter collars on a simulated animal. Individuals attempting to build their own transmitters, however, may have difficulty.

Because of the effect of an animal's body on the transmission range of a transmitter, radios should only be tested for range while attached to an animal or some simulation of an animal's body. One standard method is to attach the collar or harness around the arm or leg of a human and to position that individual so that the collar is at approximately the height above ground and the position in which it would usually be when attached to a study animal. Thus, for example, to range test a wolf collar, the collar could be attached upright around the thigh of a person sitting in a chair.

As indicated earlier, the optimum length of a transmitting antenna is a function of frequency. Decreasing its length tends to make it less efficient and thus to decrease its transmission range. The best range is usually obtained from an antenna that protrudes from the top of a collar straight up for its full length. Neither the thickness of the antenna nor the type of metal has much effect on the antenna's transmission range.

The height and elevation of an animal, of course, depend on the species. Clearly, a collar on a giraffe would transmit farther than one on a snake. However, a collar on a snake peering over a cliff might transmit farther than one on a giraffe in a valley. Radios on squirrels in the treetops or on flying birds produce signals with greater ranges than radios on the same creatures on the ground (Pienkowski 1965).

The role of soil type, topography, and weather in transmitting range has not been measured. Certain types of soils may allow better transmission above them. Sparse or dry vegetation allows better signal transmission than dense, wet vegetation, although such differences have not yet been quantified. A special radio-tracking system for overcoming some of the problems in dense subtropical bush has been devised (Anderson and De Moor 1971).

The sensitivity of various commercially available receivers varies somewhat, making it a critical factor in determining the range of transmission. Regardless of price, it is most economical to buy the most sensitive receiver. This is especially true if most of the tracking is to be done from aircraft, since searching time is extremely expensive. Ideally, a researcher should test and compare several different commercial receivers. In such tests, all other variables must be held constant so that only the receivers are tested.

One of the most important factors determining the range of a transmitter is the receiving antenna, which has several characteristics influencing range. A Yagi antenna gives more range than a loop. The H antenna gives less range than the Yagi but more than the loop. Regardless of the antenna type, the most important factor influencing range is the elevation of the antenna. Within certain limits, the range of the receiving antenna increases by about 50% for each doubling of height above ground (Cochran 1980), except that the first 6 to 10 ft (2 to 3 m) above ground yields the greatest increase because of the strong negative effect of the ground itself. Therefore, it is better to stand on top of a vehicle with a receiving antenna than to use it from the ground. Obviously, an antenna attached to an aircraft gives far more range than one on the ground. Searching for signals from mountain tops gives greater range than searching from valleys, and tower-mounted antennas give greater range than hand-held antennas.

When a researcher is studying a species having a rather limited range, for example, many herbivores, or is interested only in individuals of any species when they occupy a certain limited study area, a few strategically placed antenna towers or masts mounted on the highest hills in the study area often yield adequate signals. Usually, an antenna lead is left dangling at the bottom of the mast, and the receiver is carried to the base each time a bearing is sought. The mast should be mounted so that it can be turned 360° by the operator, or the antenna should be mounted on a rotatable apparatus operated from the bottom of the tower.

The most commonly used antenna is the Yagi, and its range depends on the number of elements, within certain limits. Furthermore, two or more antennas can be combined one or two wave lengths apart in such a way as to increase their directivity and thus their range (Figure 18). Such an arrangement is called "stacking" the antennas (p. 52).

RANGE TESTING

There are two types of range testing one should perform before beginning a radio-tracking project. First, a standard test range should be set up to allow comparison of various transmitters. The testing should be well planned and thought out so that the testing conditions can be held constant throughout the entire duration of the project. Each transmitter tested should be in its complete package (collar, harness, etc.) and should be placed around a human's leg or arm or in his or her hand to simulate its attachment on an animal. This step is not absolutely necessary but will yield a finer comparison among transmitters. The transmitter to be tested should be placed in the same spot each time and oriented the same way; for example, a collar should always face the same direction and be placed the same height above the

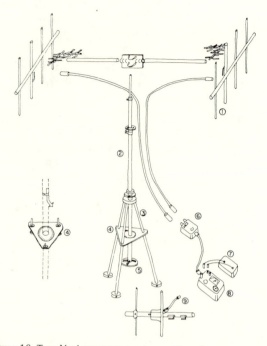

Figure 18. Two Yagi antennas can be positioned one or two wave lengths apart to increase the signal pickup of the system: (1) Yagi four-element Cushcraft antenna; (2) 9.1-m (30-ft) push-up mast; (3) 1.5-m (5-ft) tripod; (4) dial and pointer platform; (5) wheel to rotate antenna; (6) peak-null box; (7) external battery (12 v); (8) receiver, which can be used with mast or portable two-element, hand-held antenna. (Illustration courtesy of the *Journal of Mammalogy* [Banks et al. 1975].)

ground, ideally the height at which an animal would carry it.

It is useful to set up the test range along a road so that the receiver can be moved quickly to various locations. The operator moves various distances from the transmitter and, using the same receiver and antenna each time, continues moving farther and farther from the transmitter until the signal can barely be detected. Such a test range in a wooded area might yield different results with the same transmitter during different seasons because of the changes in vegetation that can cause varying degrees of signal attenuation. Nevertheless, this type of range allows valid comparisons to be made among transmitters tested at the same time.

The testing of transmitters on a range allows the researcher to discard or return transmitters with very poor range. Second, when it is known which transmitters have the greatest range, those transmitters can be reserved for use on individual animals that might be the farthest ranging, for example, those expected to disperse rather than those considered resident.

The second type of range testing involves placing the transmitter in various parts of the study area, ideally on a human's arm or leg in a position comparable to the way an animal might carry it. Then, by moving around the study area to different points, for example, mountaintops, valleys, and hillsides, the researchers can determine the workable range of the transmitter. When several types of tracking systems are to be used, the ranges of each should be tested on transmitters placed in various areas. Only the transmitters giving best and worst range need to be tested in this way.

Transmitter Life

The life of the transmitter depends on several factors. Its ultimate life is a direct function of the weight and type of batteries powering it. For example, a 50-g lithium cell will power most transmitters for about 500 days, assuming a commonly used current drain of about 0.3 to 0.5 mamp. On the other hand, a solar cell, theoretically, would power a transmitter forever as long as the sun was shining.

A second class of factors determining transmitter life are the electrical properties. For example, the more current drain when the transmitter is operating, the shorter the life, and vice versa, except that below a certain minimum current drain the transmitter will not operate. A transmitter requiring 2.7 v to operate will last half as long as one requiring 1.35 v yet having the same weight battery.

Another factor affecting transmitter life is the protection afforded it. If the protection is perfect and the electronic components and connections are all flawless, then the transmitter should function as long as the battery lasts, which depends on the battery type and weight. Transmitters should be protected both from the animal and from the animal's environment. The transmitter package should be waterproof and impervious to scratching and biting. Protection is usually accomplished with such "potting" compounds as acrylic (Mech et al. 1965), epoxy, fiberglass, or electrical resin (Kuechle 1982), silastic (Donaldson 1980), or Plasti-Dip (Jansen 1982). Some companies mount their transmitters in small metal cans, evacuate the air in a special chamber, and seal the can. The transmitter leads to the antenna and battery are potted along with the battery. The whole package is then mounted on a collar or harness made of one of various types of materials.

Attachment Techniques

The most common method of attaching transmitters to animals is by a collar (Figure 19). With some species the collar can be slipped over the head, but for others it must be opened around the animal's neck, adjusted precisely, and then bolted shut. Several considerations should be kept in mind when attaching a radio collar. Room must be left for growth and/or seasonal enlargement of the neck. However, the collar must be snug enough that the animal cannot get its front foot through it. I know of one cottontail rabbit (*Sylvilagus floridanus*) and two deer (*Odocoileus virginianus*) that put their feet through the collar and could not extract them.

Figure 19. A snowshoe hare (*Lepus americanus*) wearing a radio collar feeds on woody browse.

Furthermore, if the collar is too loose, an animal may be able to pull it off over its head. Mammals with large ears are less likely to do this. On species with long, tapering necks, for example, deer, the collar should be placed tightly enough so that it does not slide up and down the animal's neck and wear off the hair.

Obviously, the less a collar affects an animal, the more accurate the results of the study (p. 79). In this respect, the weight of the entire package is important. The best rule of thumb regarding weight is to make the total package as light as possible while still allowing the package to perform well enough to provide data for the study. With some species only trial and error can provide a guide.

Species with short, thick necks can pull a collar off. A harness is often useful for them. Such a harness involves a collar and other straps encircling the chest of the animal and connecting to the collar, usually over the back. Harnesses must be made of materials that the animal cannot chew, because most animals can reach the chest part of the harness with their teeth. On beavers (*Castor canadensis*), a collar encircling the base of the animal's tail has been used successfully (R. Buech, personal communication, 1982). For species that are especially difficult to collar or harness, such as river otters (*Lutra canadensis*), deer mice (*Peromyscus leucopus*), and mink (*Mustela vison*), tracking transmitters have been implanted into the abdomen (Smith and Whitney 1977; Melquist and Hornocker 1979; Smith 1980; Eagle, Norris, and Kneeble, personal communication, 1983). With bears, transmitters mounted on ear clips have been used (Servheen et al. 1981). Obviously, such transmitters can use only a small battery or a solar cell.

For birds, transmitters have been mounted on harnesses (Nicholls and Warner 1968, 1972), tail clips (Bray and Corner 1972), and wings (Figure 20). For short periods, less than a month, for example, small transmitters can be glued to the back of the bird (Figure 1) after a few feathers have been cut off just above the skin (Raim 1978). An excellent list of references for radio-attachment techniques for birds has been published by Cochran (1980: table 29.1).

Transmitter Reliability

The reliability of a transmitter varies, depending on whether it was homemade or commercially made, which company produced it, and on what species it is used. In general, privately made transmitters are not as reliable as

Figure 20. A solar transmitter has been attached to the left wing of this endangered California condor (*Gymnogyps californianus*). (Photo courtesy of Helen Snyder, Condor Research Center.)

commercial units. The latter have the advantage of professional construction, manufacturer's experience, testing with expensive and sophisticated equipment, and partial guarantees. The biggest advantage of privately made transmitters is that they have the individual attention of the person building them at all stages of construction and testing.

As with any other product, the quality of the radio transmitter varies with the manufacturer, and for a given company, with time. Furthermore, some manufacturers may make better transmitters for certain applications, while other companies may make better ones for others. The best way to determine which company's product is most reliable for a specific application is to check with other people who currently are radio-tracking the species to be studied. Reliability is the most important characteristic of a radio collar. No matter how long it is supposed to last or how far it should transmit, when it fails prematurely, its other characteristics become meaningless.

Transmitter reliability also depends considerably on the species being studied. Large carnivores such as leopards (*Leo pardus*) and wolves seem to be quite rough on collars, perhaps because when fighting they can grab an opponent's collar by fang or claw and damage it. On the other hand, it is relatively easy to keep transmitters working on large ungulates.

Probably the most important factor determining the reliability of a transmitter is the protection given its components. Transmitters should be completely waterproof, should be able to withstand great temperature extremes both electrically and physically, and should be able to withstand banging on rocks, chewing, biting, and scratching. Critical components of the transmitting system should be particularly well protected, for example, turn-on leads, antennas, and battery leads. One small crack or disconnection, or wear or corrosion in the right place, can interrupt transmission. Turn-on leads should be soldered together carefully. A hot soldering iron should be touched to the lead until solder flows onto it. After both leads are so "tinned," they should be twisted together and heated again with the soldering iron until the solder flows. Then, they should be covered with dental acrylic or epoxy.

Premature failure of a transmitter is a disconcerting problem for the radio-tracker. When a signal suddenly seems lost, there are several possible explanations: (1) premature failure of the transmitter, (2) movement of the animal out of the receiving system's tracking range, (3) a faulty receiving system, or (4) capture of the animal by a human being and subsequent destruction of the transmitter.

It is difficult, and often impossible, to separate the possible causes of signal loss. Some tests can be made, however, and some inferences drawn from various types of evidence. If a transmitter's signal had a history of being intermittent and suddenly that transmitter is not heard at all, it is reasonable to assume that whatever caused the intermittence progressed and turned off the transmitter. A sudden change in pulse rate or any other characteristic of a signal also might indicate impending failure. Therefore, every time an aberrant signal is heard from a study animal, a note should be made. A spare transmitter can be used to test a possible problem with the receiving system.

Since many individual animals may suddenly disperse after having been tracked for long periods, suspected dispersal out of range should be checked by using long-distance tracking systems such as mobile land-based systems or aerial systems. A search for an animal with a lost signal should be conducted as soon as possible after the loss is noticed. A dispersing or migrating individual may continue to move farther away as time passes. Every day one waits before attempting to find a lost signal gives the individual a better chance to escape the detection radius. There is clearly a great advantage in knowing the abilities of study species to travel long distances in short periods.

When all other factors can be ruled out, premature transmitter failure should be suspected, and a note to that effect should be made in the animal's records. Then, if that individual is ever recaptured, the transmitter can be tested and returned to the manufacturer with a request for an autopsy. One should provide the manufacturer with all the pertinent details about when the transmitter was started, when it was put on the animal, how long it lasted, and when the failure was noted. A definite request should be made to determine the cause of failure, and one should continue to follow up on this request until information is provided. This is the only way the manufacturer and the researcher can know the causes of failures and thus improve their equipment.

One further note here is important. Because many animals are harvested by human beings, it is not unusual for a person to kill a collared animal. For this reason, a

Figure 21. A small tag giving the name and phone number of the researcher and embedded in the radio package improves chances for retrieval of the transmitter and information about the study animal even after the radio has stopped transmitting.

small but conspicuous message should be placed on the radio package giving the animal's number, agency's address, and researcher's phone number (Figure 21). A call from the person who retrieved a collar can often save the researcher valuable search time and satisfy his or her curiosity about what happened to the animal's signal. If a reward is offered for information and the return of the radio, a notice stating that should be included in the message.

RELIABILITY TESTING

Of the relatively few transmitters that fail for electronic reasons, a high percentage occur during the first few weeks of life. Therefore, long-lived transmitters should be turned on 2 to 4 weeks before they are attached to an animal so that any defective transmitters can be discarded. A marked change in the pulse rate or a complete loss of signal indicates a faulty transmitter.

When using mercury batteries, a biologist can check

their reliability by having them X-rayed (Harding et al. 1976), or he or she could ask the manufacturer to do so. In the latter case, the biologist might have to assume the costs of the discarded batteries, but the payoff in reliability would be well worth it.

To Buy or To Build?

As indicated earlier, there are several advantages to buying radio-tracking equipment from commercial manufacturers. It is sometimes difficult to find compatible components for transmitters, but commercial manufacturers can afford to buy such components in large quantities, test them for compatibility with each other, and discard those that are incompatible. Manufacturers are also better able to test the components, and they usually keep informed about new types of components with better characteristics. Their general experience, specific knowledge about radio-tracking equipment, and practice generally ensure the higher quality and efficiency of commercial transmitters.

In addition, most of the manufacturers of radio-tracking equipment have developed transmitter circuitry beyond the level of the circuitry for which information has been published (p. 18). Furthermore, most companies offer some sort of guarantee on their equipment. Obviously, a guarantee cannot bring back a valuable animal on which a transmitter suddenly failed, but it does help somewhat, especially if transmitters are tested before field application and failures are noted then.

The main advantage of making one's own transmitters, assuming that the necessary time, equipment, and knowledge are available, is that the parts are much cheaper; if the labor is available, more transmitters can be built than could be purchased for the same amount of money. Second, one can custom build each transmitter, taking as much care as possible. Third, one might be able to overcome the long lag often encountered between the ordering of equipment and its delivery. Researchers interested in building their own equipment should consult Zimmerman and others (1975).

Nevertheless, when all of the advantages and disadvantages of buying and making telemetric transmitters have been considered, most people conclude that it is better to buy the equipment from a reliable company (Appendix III) than to make it themselves.

Types of Receiving Systems

The basic receiving system consists of a radio receiver, an antenna, and a recorder, which usually is a human being. This basic system can vary in several ways, however. The various types of antennas have already been mentioned, and several makes and models of receivers are commercially available. This section describes the different ways in which receiving systems can be set up physically and used.

PORTABLE TRACKING SYSTEM

In the simplest version of a radio-tracking system, the basic components are carried by the operator. The receiver is usually placed in a carrying case or sidepack, which can be hung from the side or front of the operator, who holds the antenna (usually a directional one) with one hand and operates the receiver's controls with the other (Figure 22). This portable system is very adaptable and allows the operator to approach an animal or even to climb a hill or a tree in order to gain better range. In areas accessible only through foot travel, the operator can scan a large area by walking

Figure 22. The simplest receiving system consists of a hand-held directional antenna, a receiver, and earphones.

wherever he or she can. The main disadvantages are that such a portable tracking system is slow and inefficient for locating animals and that disturbing the animals is more likely. Using this type of system is necessary, however, whenever one wants to home in on an animal or to find exactly where it is denning, nesting, resting, sleeping, or lying dead.

STATIONARY RECEIVING SYSTEM

A stationary receiving system includes an antenna mounted on a tall mast, which gives the system more range. Usually, the mast is located on the highest terrain in the area. A large Yagi antenna, or a pair of them stacked (p. 39), can be mounted atop the mast. Either the mast or the antenna must be rotatable in order to allow scanning for the signal when locations are desired. (If only data on activity or the presence or absence of a signal are desired, however, the antenna need not be rotatable.) The operator or recorder can remain at the base of the mast and pick up the signal via the antenna cable.

The stationary receiving system is useful with species that move over only small areas or when the researcher is only interested in what happens on a small piece of land that the tower can monitor. The disadvantage is obvious: the area of pickup is limited. Nevertheless, a series of stationary tracking systems scattered throughout the study area on prominent elevations can supplement a portable or even a mobile tracking system, which may not give as much range.

MOBILE TRACKING SYSTEM

If there is a sufficient network of roads or trails through the study area, a mobile tracking system can be used (Verts 1963; Hallberg et al. 1974; Bray et al. 1975; Kolz and Johnson 1975; Hutton et al. 1976; Whitehouse and Steven 1977). Such a system employs a permanently mounted Yagi antenna and antenna mast on the roof of a vehicle, with the cable leading down into the vehicle, where the operator monitors the signal through the receiver (Figure 23). This type of system can permit scanning large areas on both sides of any road the vehicle can traverse.

Some mobile tracking systems use two Yagis "stacked" side by side (Figure 18). Stacking antennas one or two

Figure 23. Most mobile tracking systems include one or two Yagi antennas mounted on the roof of a vehicle.

Figure 24. Mobile tracking systems can also be carried on elephantback.

wave lengths apart increases the "gain," or receivable signal strength, to the receiving system. With an appropriate switch, called a null/peak box, a stacked antenna system also gives greater accuracy. The switch allows a person either to integrate the signals from the two antennas into a peak or to more or less subtract them, giving a null. The null is a very sharp, V-shaped area of no signal that emanates from the antennas. The peak reception pattern, which is stronger, is first used to detect the signal. Then, the weaker, but more accurate, null pattern is used to determine the precise bearing to the transmitter. Commercial suppliers of the null/peak system provide detailed directions for its use.

The antennas are attached via a crossbeam to a telescoping mast. The mast can be rotated by the operator inside the vehicle. A compass and a pointer inside the vehicle, indicating the direction the antennas are pointing, enable the operator to take compass bearings to the signal. In terrain where vehicles cannot go, portable tracking systems can be carried on horseback or elephantback to excellent advantage (Figure 24).

AERIAL TRACKING SYSTEM

Aerial radio-tracking systems use antennas attached to the wing struts of fixed-wing aircraft or to various parts (for example, skid struts) on helicopters (Mech 1974; Whitehouse and Steven 1977; Gilmer et al. 1981; Inglis 1981). For such a system it is important to keep at least the side and front of the antennas as far as possible from any other metal in the aircraft. This is usually done with extending pipes to which the antenna beam is attached, often with hose clamps (Figure 25). The antennas can be used as they point forward (that is, so that the strongest signal is received when the aircraft is heading directly toward the animal) or sideways (that is, so the strongest signal is coming from the side of the aircraft) (Figure 26). Both methods have advantages, but most people tend to use the "side-looking" system. It is most important that the antennas, methods of mount, and distance from parts of the aircraft are the same on each side of the plane. The reason for this will be covered in the discussion of the aerial homing technique.

It is also very important that the antennas be mounted securely to the aircraft so that they do not interfere with the flight. Several methods are used. The interested researcher should consult colleagues who have used the aerial tracking system to obtain details about the type of mounts. In the United States and probably in other countries, one may want to seek the Federal Aviation Agency's (FAA's) approval of the clamps used to attach the antennas to the aircraft. With FAA-approved clamps, a biologist can usually hire an aircraft for radio-tracking from almost any airport on short notice.

Usually, Yagi or H-type antennas, rather than dipole antennas, are used in aerial tracking, because they have much greater gain and thus more pickup range. Furthermore, they allow more effective tracking, because they are directional. A left/right switch box is also required. This allows the operator to switch between the two

A

B

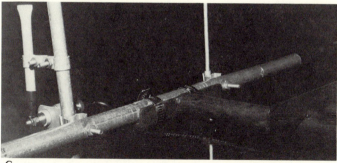

C

Figure 25. (A) Yagi antennas are attached to aircraft wing struts via custom-made devices held on with standard automotive hose clamps. The antennas should extend in front of the wing and as far from the metal as possible. (B) The close-up shows one type of attachment device for holding the base of a hollow pipe to which the antenna is attached at other end. (C) The close-up shows one type of angle-iron device, welded to the antenna end of the pipe, to which the antenna is attached.

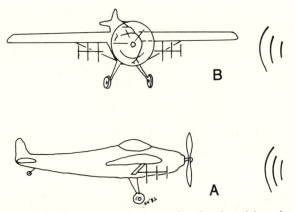

Figure 26. Yagi antennas can be oriented so that they pick up the strongest signal from A in front of the aircraft ("forward looking") or from B the side ("side looking").

antennas in order to determine from which side of the aircraft the signal is coming. Some such boxes have a switch allowing both antennas to listen for a signal simultaneously; this is useful when one is scanning for a lost signal. Then, when the signal is found, the switch is thrown to another position, allowing each antenna to pick up the signal separately so that the operator can determine which is receiving the stronger signal.

When using both antennas to search for a distant signal, one should use a coaxial "power splitter" type of connector for the two antennas; otherwise, the imped-ance of the antenna system with both antennas con-nected probably will not match that of the receiver, and considerable range may be lost. Many such commercial connectors sold by manufacturers of equipment for wildlife telemetry do not take this problem into ac-count. If a coaxial connector is not available, maximum search range can be obtained by alternately connecting one of the antennas at a time to the receiver.

The coaxial cables from the antennas are fed through the aircraft's door or window (between its edge and the body, where there is rubber weather stripping), or on a Cessna aircraft through the wing vents, and into the cockpit, where they are attached to the switch box. The switch box is then attached to the antenna jack on the receiver.

Because of the high cost of operating aircraft, it is imperative to use the most efficient receiving equipment available. Eliminating only a few minutes from each search for an animal cuts the cost considerably and saves enough money to justify the purchase of the most efficient and reliable equipment available. For the same reason, it is always desirable to test the tracking system before taking off. A faulty system can lead to long periods of unproductive flying before one realizes that the receiving system is not working properly. Testing can be done on the ground by using a transmitter left at the airport.

The weakest parts of the aerial tracking system, and so the parts most likely to be damaged, are the points at which the cables attach to the connectors. With continued use, these begin to wear and short out. Although the background noise can still be heard, usually signals cannot be received when the connections are shorted. It helps to wrap each connector and first part of each cable with several layers of electrical tape.

To determine what part of the tracking system is malfunctioning, one simply tests the various parts. For example, the switch box can be tested by attaching one antenna lead directly to the receiver. If the signal is received but cannot be received with the antenna plugged through the switch box, then the switch box is faulty. The receiver can be tested by substituting another receiver. This process is continued until all parts of the receiving system have been checked and the problem has been solved.

Radio-Tracking Techniques

The simplest radio-tracking technique uses a portable, hand-held receiver and a directional antenna. With experience, one can determine the general direction of the radioed animal and some estimate of its distance. Obviously, this type of system is most useful with smaller, less mobile animals. All that is necessary is to tune in the signal, turn down the gain as low as possible while still keeping a signal, and then swing the directional antenna slowly in an arc. At the point when the strongest signal is heard, the antenna is pointing toward the animal. Judging the distance to the animal depends on the operator's experience with a given type of radio equipment and its use in the area.

TRIANGULATION

A much more accurate way to determine the actual location of an animal is to use triangulation. Triangulation requires taking bearings from at least two points and assuming that the radioed animal is near the point at which the bearings cross (Figure 27). Triangulating can be done in several ways. Two operators, each using a receiver and an antenna, can simultaneously take bearings to the same animal from two widely separate known locations. This requires two-way radio communication between the operators. Each records his or her own position on a map and then the compass bearing from that position to the animal. The amount of declination from true north must be considered with each bearing so that the true location of the animal can be obtained (Figure 28). With animals at rest or moving slowly, one operator can triangulate by moving rapidly between two points and taking bearings from each point (Figure 27).

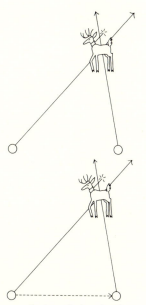

Figure 27. Two ways of triangulating: (top) bearings are taken from two points simultaneously and the points at which the bearings cross are determined, or (bottom) a bearing is taken from one point, and after the tracker moves to another point, a second bearing is taken from there.

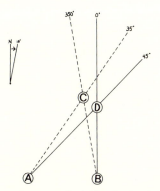

Figure 28. To account for magnetic declination in plotting points where signal bearings cross (making a "fix"), the amount of declination must be subtracted from the magnetic (compass) bearing before plotting on a map. For example, if a bearing from A is 45° and one from B is 0° and the declination is 10° east of north, then the true map bearings are 35° and 350°; therefore, the animal is actually at C rather than at D.

Triangulation errors (p. 75) result from the inaccuracy of the directional antennas and the misjudgment of the operators using them. Many directional antennas, therefore, have an accuracy of, for example, plus or minus 2°. When such latitude is considered with both bearings, the point at which they cross, or the "fix," can be different from the point at which the radioed animal actually is located. Furthermore, the more acute the angles at which the bearings cross, the greater the possible error (Heezen and Tester 1967). In addition, the farther away the operator is from the animal when the bearings are taken, the greater the error may be. The distance between the two points from which the bearings are taken ideally should be enough to allow the bearings to cross at about 90°. Finding this distance is a matter of judgment that requires considerable experience. These problems can be identified easily by placing test transmitters in known locations and trying to triangulate on them. By mapping several triangulations from different points at various distances apart and at different distances from the transmitter, one can determine the degree of error under various conditions.

HOMING

Homing is a common method of radio-tracking. It is used for such things as locating radio-tagged birds on nests, finding dead radio-tagged animals, finding transmitters that have come off an animal, locating denning and resting sites of radioed animals, finding drugged animals hit with radio darts, and locating radioed animals directly for observation. The basic homing technique employs antenna directionality, signal strength, and movement by the operator.

When homing from the ground, the best approach is as follows: (1) Tune in the signal, (2) *reduce the gain to minimize the signal,* (3) rotate the directional antenna 360° while listening to the signal, (4) determine the approximate direction from which the signal is strongest, (5) *again reduce the gain* and minimize the signal, and (6) slowly wave the antenna over the arc from which the signal is received and again try to determine the precise direction from which the signal is strongest. (Earphones are of considerable value to block extraneous noise, especially wind.) (7) Next, move closer to the signal while slowly waving the antenna over an arc of about 90°. (8) As the signal is approached, continue to judge the direction from which the signal is maximum, and (9) *continue to reduce the gain.* (10) If the signal is lost after a certain distance, do not turn up the gain; rather, return to a point at which the signal was heard before and move in a different direction. (11) If unsure of the direction, scan a full 360° and listen again for the peak signal; or from the last point at which there was a signal, make test moves in four directions 90° apart and listen for an increasing signal, which can only be heard when the operator walks in one of the four directions. (12) When the gain control has finally been adjusted to its lowest position and the signal can still be heard, the animal or the transmitter is usually very close. If the transmitter has not yet been found, pinpoint by removing the antenna from the receiver, which will eliminate the signal, and then turn the gain back up to the point at which the signal can barely be heard. You must then rely entirely on the change in signal strength resulting from your moving back and forth a few meters. Through trial and error and by continually returning to points at which the signal is strongest, you should find the transmitter. Remember, however, that the faster you move while homing, the easier it is to tell when the signal strength is changing and, therefore, when you are getting closer to the transmitter.

AERIAL RADIO-TRACKING

Aerial radio-tracking is basically homing in an airplane. The following directions apply to the side-looking antenna arrangement discussed earlier. To begin aerial homing, one should direct the pilot to begin circling above the animal's last known location and spiraling upward. The operator then tunes for the signal, alternating between the two antennas. The plane should continue to soar as high as necessary to pick up the signal, which with far-ranging species may require altitudes of up to about 10,000 ft (3000 m). When the signal has been detected, the plane should continue circling through at least another 360° to ensure that the operator knows the direction from which the loudest signal has come. Again, keeping the gain minimal makes it easier to detect this point.

The next step is to head the plane toward the signal (Figure 29). The signal may then disappear, because the antennas will not be pointing toward it but rather will be pointing 90° from the signal's path. However, eventually as the plane approaches the transmitter, the signal will be detected from the right or left antenna. Unless the plane is heading precisely on the signal's path, which is rare, the signal will first be picked up from one side of the aircraft, not from both (Figure 29). Therefore, the operator should continue to switch from one antenna to the other while listening for the signal. If the operator does not hear the signal in enough time, he or she can always direct that the plane be turned 90° temporarily, at which time one of the antennas will be pointing toward the signal and the signal can then be picked up again either ahead or behind the turning point. If for some reason the signal is not heard again, then another circle should be made until the signal is picked up. As a last resort, one can return to the area where the signal was heard originally and double-check the signal's direction.

As the plane approaches the radioed animal, signals will be received from the front of each antenna; at that time, the operator should switch from one antenna to the other to detect the side on which the signal is stronger. When this has been determined, the operator signals the pilot to turn 30° to 45° toward the stronger signal (Figure 29). Meanwhile, the operator continues to switch back and forth between the two antennas and

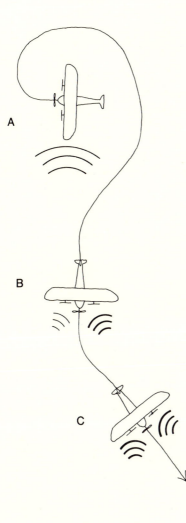

Figure 29. With side-looking antennas (Figure 26), the basic approach to homing from the air is to (A) circle while listening from the antenna on the outer wing until the signal is heard, (B) head in the general direction of the signal, and (C) angle the aircraft toward whichever side the signal is stronger until signals from both sides are about equal. The aircraft should then continue on straight toward the signal.

to direct the pilot until the signal is of equal strength through both antennas. At this point, the aircraft should be heading directly toward the radioed animal.

Throughout this process, at least when the signal is heard from the front of the antennas, the operator can direct the pilot to descend gradually. If the signal is lost as the altitude is decreased, the pilot should be notified to level off until the signal is received again and then gradually to descend again. Minimizing the aircraft's elevation enables the operator to make finer distinctions in deciding where the radioed animal is, because the signal projects from the animal in a cone-shaped pattern, with the larger end of the cone higher up (Figure 30).

When the signal is heard at about equal strength through both antennas, the operator should begin to turn down the gain control in order to minimize the signal. The signal, however, will increase again as the plane continues to approach the radioed animal. At some point, which can only be determined through the judgment and experience of the operator with individual radios, the gain control will be at some low point, where the operator can then leave it and begin concentrating on when the signal peaks.

At this time, the signal-strength meter becomes very useful, for one can see the needle swing as it gradually increases in amplitude and then eventually begins decreasing. Because of environmental conditions, the signal fluctuates as it increases so that one must watch the overall pattern rather than individual movements of the needle. As the aircraft passes over the radioed animal, the signal becomes strongest, and as the plane continues on by, the signal eventually begins to decrease. The instant the operator detects the decrease, he or she should direct the pilot to make a 180° turn, and he or she should listen again for a peak (Figure 31).

At this time, it is best to listen to the signal from one antenna or the other rather than to switch back and forth. Even though the signal would be about equal from both antennas when the plane is pointed directly

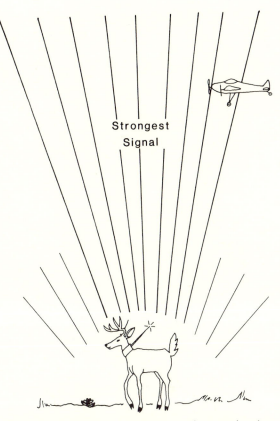

Figure 30. The signal from a transmitter near the ground projects into the atmosphere in a cone-shaped pattern. Therefore, from an aircraft it is easier to distinguish various signal strengths at lower altitudes than at higher ones.

at the radioed animal, chances are good that the animal is actually located to one side of the aircraft or the other. The closer the plane gets to the animal, the clearer it becomes that the signal is louder from the left or right antenna. When the operator has determined which signal is louder by switching back and forth, he or she then can start listening on the stronger side only.

Therefore, when the operator ascertains, via the method described above, that the plane is about to approach the radioed animal again, he or she should

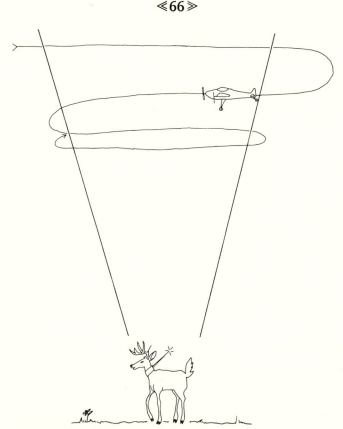

Figure 31. In aerial homing, once the signal reaches its maximum and begins to diminish, the aircraft has passed the radioed animal and must turn 180° and decrease its altitude so that the tracker can distinguish more finely the animal's location.

direct the pilot to circle to the direction from which the signal is stronger (Figure 32). If the animal is within the circle, the signal should always be stronger from the antenna pointing toward the center of the circle. Thus, as the plane circles, the operator should switch back and forth between antennas. If the signal is always stronger from the inner antenna than it is from the outer, then the animal is inside the circle (Figure 32).

If, however, at one point the signal is stronger from

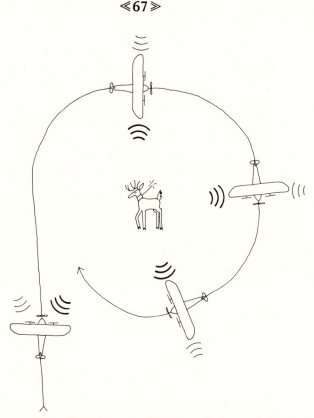

Figure 32. When the aerial tracker thinks he or she knows about where the animal is, the pilot should circle around the point while the tracker switches alternately between the inner and the outer antennas. If the tracker's judgment is correct, the signal will be stronger from the inner antenna during the entire 360° of the circle.

the outer antenna than from the inner, this means that the plane is circling to one side of the animal. Either the plane must then circle in the opposite direction when the stronger signal is heard from the outer antenna, or the entire circle must be moved in the direction of the outer signal (Figure 33). The circling and switching process should be repeated until it is clear that the signal is always stronger from the inner antenna.

With a small aircraft, such as a Supercub or an Indian

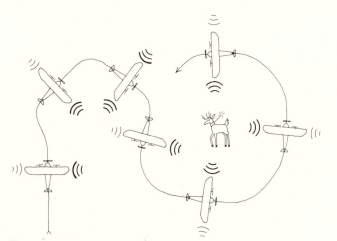

Figure 33. If, at one point in the circle around the hypothesized location of the animal, the signal from the outer antenna becomes stronger, then the radioed animal is not within the circle. The aircraft should circle in the opposite direction around the new supposed location of the animal, and the test (Figure 32) should be repeated there.

Bushbuck, the circles can be small enough so that, once the radioed animal is found to be within a certain circle, the location is precise enough that the animal can be observed if it is not hidden by ground cover. With larger aircraft that make much larger circles, however, it is sometimes necessary to pinpoint the radioed animal more closely. This can be done by making a pass through the center of the circle and switching from one antenna to another to determine in which half of the circle the radioed animal is. A second pass made across the circle perpendicular to the first narrows the location to the nearest quadrant (Figure 34). This technique can be repeated within the quadrant to narrow down the location even further. It is sometimes necessary to turn the gain down even farther when making these kinds of flights so that it is easier to distinguish on which side of the aircraft the signal is stronger.

During aerial homing, it is sometimes difficult to tell

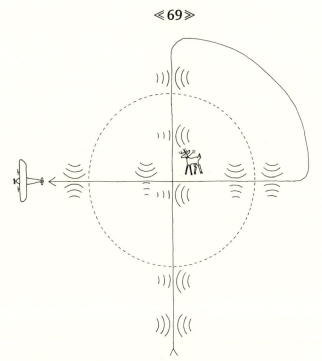

Figure 34. If the tracker knows the approximate location of the radioed animal and wants to try to observe it from the air or to determine its location even more precisely, the "fly-by" method of pinpointing can be used. The aircraft flies low across the center of the circle in which the animal is known to be, while the tracker switches rapidly between right and left antennas. The animal is in the semicircle from which the stronger signal is received. Repeating the procedure perpendicularly to the first flight path enables the tracker to determine in just which quadrant the animal is located. The procedure can be repeated until the animal's precise location is learned.

whether the aircraft is still approaching the radioed animal or has passed it. One way of distinguishing is to turn 90° toward where the operator thinks the animal is located. While turning, the operator switches to the outer antenna to see whether the signal seems louder from it or from the inner one. In other words, if the aircraft turns right, the operator listens from the left antenna; if the signal is stronger on the left than on

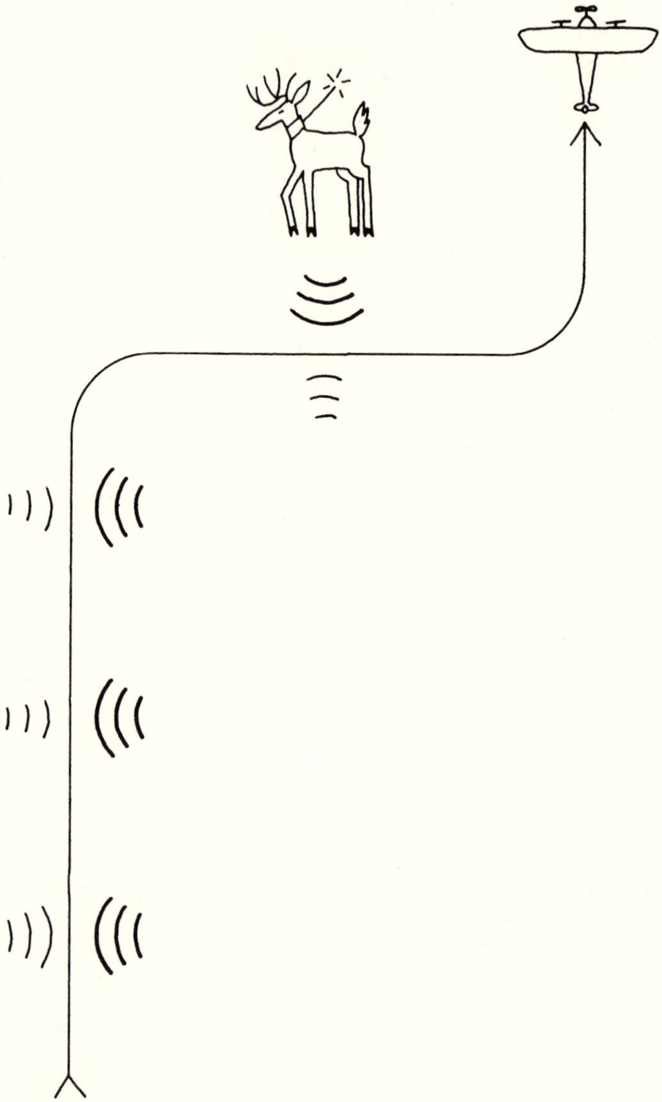

Figure 35. As the aircraft approaches an increasing signal when the tracker is trying to determine just when the aircraft is opposite the animal, he or she must use trial and error. When the signal seems at its strongest, the aircraft should make a 90° turn toward it, while the tracker switches between antennas. If the plane turned right and the stronger signal came from the left antenna, the animal would be still farther ahead along the original flight path; otherwise, it would just have been passed.

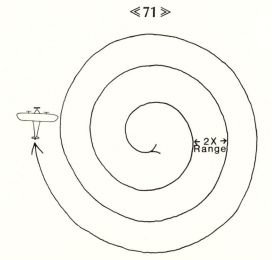

Figure 36. This aerial search pattern should be followed for an animal that dispersed in an unknown direction. Start at the animal's last known point and spiral outward, with the distance between parallel flight paths being no more than twice the known minimum reliable range of the transmitter to the aircraft for a given altitude.

the right, he or she knows the aircraft has turned too soon and that the radioed animal is still ahead along the original flight path (Figure 35). If, however, the signal is stronger from the right antenna, the aircraft has already passed the animal and must make another right turn to head back toward the animal.

AERIAL SEARCHES

For far-ranging animals or those that disperse long distances, it is sometimes necessary to search large areas, especially when considerable time has elapsed between radio-tracking attempts. The most efficient procedure in such cases is to fly directly to the animal's last known location and to listen for the signal on the way. If the animal is not there, the operator should begin to spiral outward and upward from that point. The flight path of this search pattern should parallel itself at a distance equal to twice the known minimum range of the target transmitter to the receiving system (Figure 36). This ensures that the maximum area is covered as quickly as possible. The operator should scan with both antennas either by constantly switching back and forth from right to left or, if the switch box is so equipped, by using the "both" mode to listen to both sides simultaneously. For best range, one should search from altitudes of about 10,000 ft (3000 m).

ACTIVITY AND PRESENCE OR ABSENCE

It may be important for many reasons to know when an animal is present or absent at a particular site. How long does a brooding bird sit on a nest? When does it begin and when does it stop? How many trips does a certain kind of mammal make back to a den to feed its young? How often do particular individuals scavenge at a garbage dump or at some other concentrated food source? These presence-or-absence questions can often be answered through radio-tracking, especially through automatic monitoring of radio-tagged individuals around the target points (Figure 17).

Such studies can be done by constant monitoring or sampling directly by a human radio-tracker. However, where human labor is expensive, far more data can be obtained through the use of automatic monitoring stations (Kjos and Cochran 1970; Gilmer et al. 1971; Harrington and Mech 1982). Such a station consists of an antenna mounted near the target area (nest, den, garbage dump), a weatherproofed box, a receiver, a power source, and a recorder (p. 31). The power source can be an automobile battery, a motorcycle battery, or a boat battery, any of which usually can power a radio-tracking receiver for 1 to 2 weeks and can be recharged. For best results, batteries made for frequent discharge and recharging, such as boat batteries, should be used.

Strip-chart recorders, event recorders, and even computers can be used with such systems. Manufacturers of radio-tracking equipment can provide further information on the types of recorders available.

In cases in which more than one radio-tagged animal

is visiting the target site, a special scanning receiver can be used. The scanner can be programmed to listen for one frequency for a given period, for example, 5 minutes, and then switch to a second, third, fourth, fifth, and so on (Figure 37).

When such a station is being set up, it is wise to calibrate it by taking a transmitter of the type on the target animal and moving it by hand around the station while a second person tunes the receiver and recorder so that the signal is only recorded when the transmitter is precisely at the point at which the animal's presence is to be recorded. Adjustment of the antenna's height and placement helps in this respect, and two-way radio contact between the person moving the transmitter and the person tuning the system is essential.

When a scanner is being used to monitor more than one animal, it is helpful to leave the calibration transmitter in the vicinity of the monitor as a constant test of whether the system is working. In addition, this calibration transmitter can be used to help identify the specific channels on which various target animals are coming in. For example, if the scan rate is 5 minutes per animal and three animals are being monitored, along with the calibration transmitter, the scanner should be set to receive the calibration transmitter, then animal number 1, number 2, and number 3 and then the calibration transmitter again, and so on. In this way, the identity of each animal on the recording can be determined by its position in relation to the recording of the calibration transmitter. In addition, the calibration transmitter recording provides a measure of time; in this example, each full cycle takes 20 minutes.

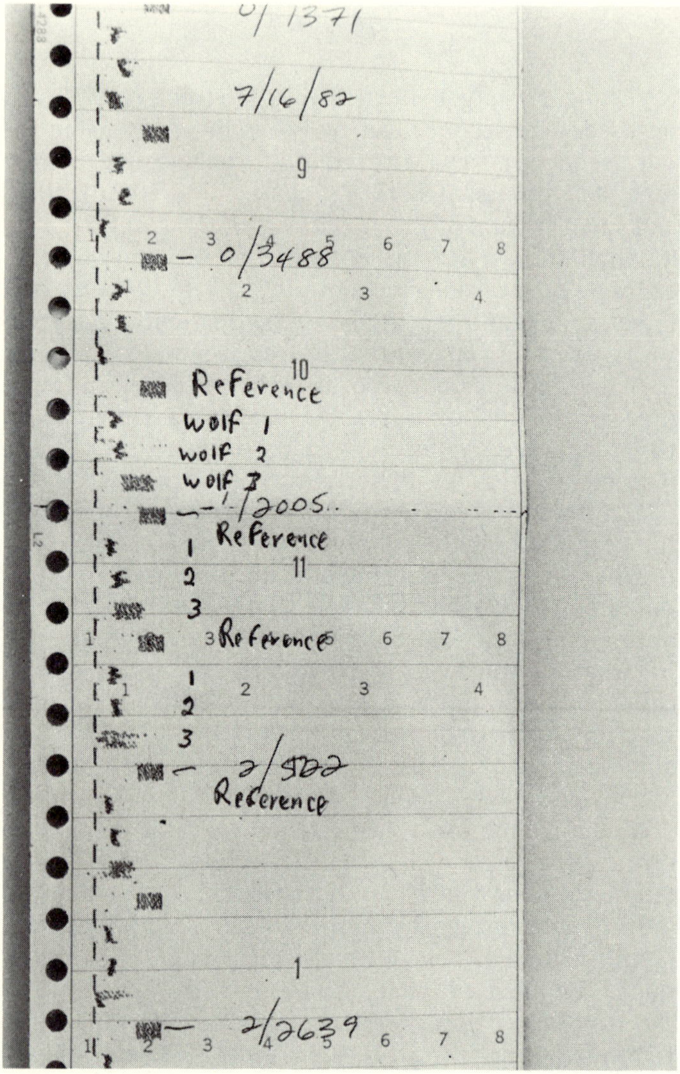

Figure 37. This example of a strip-chart recording shows radioed wolf visits near a recording station. The regularly spaced blocks represent signals from a reference transmitter placed nearby to test the system constantly and to provide a time indicator. The ragged recordings between reference signals indicate the presence of wolves moving around nearby, and the thin, flat marks interspersed at the bottom are blank frequencies used for comparison.

Errors and Accuracy

As indicated earlier, much radio-tracking is done by triangulation during which bearings to the transmitter are obtained from two locations. After consideration has been made for the degree of magnetic declination in a specific area, the intersection of the two bearings represents the best estimate of the location of the signal's source. However, no antenna is precise in its directionality; that is, there may be as much as 2° or 3° of latitude in determining the bearing of a signal. The same is true when the bearing is taken from the second point. Therefore, the actual location of the animal may be anywhere within a diamond-shaped polygon formed by the intersections of the two sets of innermost and outermost possible bearings from the two locations (Figure 38).

The precise shape and area of the "error polygon" is determined by the following: (1) the accuracy of the directional antenna, (2) the distance between the two receiving points, (3) the distance of the transmitter from the receiving points, and (4) the angle of the transmitter from the receiving points (Figure 39). The closer together the two receiving points are, the longer the error polygon is. The more the angle varies from 90° between one receiving point, the signal's source, and the second receiving point, the larger the error polygon (Figure 39). It should be apparent that the most accurate estimate of a radioed animal's location is based on the receiving points that are closest to the animal and at 90° from each other (Figure 39).

One method of overcoming triangulation error is to take three bearings instead of two and to estimate the

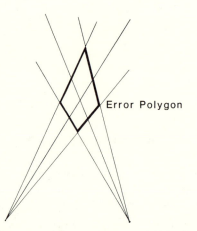

Figure 38. Because receiving antennas may yield bearings that vary by 2° or 3°, the actual location of an animal may be different from the apparent location. During triangulation, both bearings may vary, forming an error polygon surrounding the animal (Heezen and Tester 1967). If the animal were in the center of the polygon, the combination of bearings might falsely indicate that it is anywhere in the polygon.

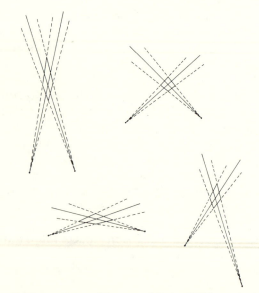

Figure 39. Error polygons (Heezen and Tester 1967) vary in size and shape, depending on the distance and angle between the two points from which bearings are taken.

Figure 40. To help overcome the problem of triangulation error, the tracker can take bearings from three points; the best estimate of the animal's location then falls in the center of the resulting triangle.

animal's location as the center of the intersection of these three bearings (Figure 40).

A second type of error common in radio-tracking is map error. Maps themselves vary in their degree of accuracy, and large-scale maps generally are more accurate than small-scale maps. Even if the map is extremely accurate, however, plotting error is another possible problem, especially on small-scale maps. On very small scale maps, the width of the lead in the plotting pencil may represent as much as a mile, so any error involved in drawing the bearings on the map could be substantial.

Another type of error comes from signal bounce. Signal bounce results when a signal hits mountains, ridges, large trees, rocks, or other natural features that reflect it. Signal bounce is characteristic of high-frequency signals, from about 100 MHz and above under most types of field conditions. Signal bounce will mislead one into concluding that a signal is emanating from wherever the bounce point is. It can often result in errors of many miles or kilometers, particularly in mountainous regions. Anyone planning to radio-track animals in such areas should first place transmitters in the study area in various positions and try to locate them from afar in order to determine the particular areas where signal bounce is most common and then to practice overcoming the problem.

The best way to overcome signal bounce is to take bearings from several different points. When the signal is bouncing, not all of the bearings will seem to come from the same point, thus giving a clue that there is some bounce. When, however, all signals seem to be emanating from the same point, chances are good that the animal is actually there.

In actual homing from the ground, rather than triangulation, any errors are corrected as the tracker proceeds toward the source of the signal and the animal, or a den, bed, or nest that confirms its location, is finally observed visually. The same may be true in aerial homing when an actual observation is made. However, if the target animal is not observed, then errors of up to 75 m in estimating the location from an aircraft are likely (Hoskinson 1976).

BIASES

Probably every investigational technique is subject to biases, and radio-tracking is no exception. Biases in radio-tracking data can result from biases in the particular times when data are sought. For example, when animals are radio-tracked only during the day, the data will be biased toward daily locations (Smith et al. 1981). Or, when an animal is tracked every day for a month and then once each week thereafter, the data will be biased toward where the animal spent the month when daily locations were obtained. These are obvious sources of bias.

Less obvious are biases resulting from relatively low success rates in finding target animals. For example, when a person is tracking from a vehicle and only finds the target animals 80% of the time, the data will be biased toward locations nearest the road; during the other 20% of the time, the animals could be out of range of the tracking system, that is, farther away from the road. Because of the possibility of this type of bias, it is wise to persist each time one tries to find an animal or to have some method of tracking on which one can rely when the usual technique fails. For example, with a wide-ranging species, one could resort to using an airplane after other methods have failed.

ADVERSE EFFECTS OF RADIO-TRACKING

There are several ways in which radio-tracking might adversely affect an animal and thus spoil the data collected. For example, a heavy radio package could hinder an animal's movement. Birds are especially subject to this problem. Furthermore, the capture itself, including any drugging or handling effects, could seriously bias data. Last is the possibility that the radio-tracker's presence could disturb the target animals. These possibilities should always be considered in designing and conducting radio-tracking studies.

The potential effects of capturing, handling, and radio-tagging animals can be assessed in a few ways. First, whenever a radio-tagged animal is recaptured, any wearing of the radio package on the animal should be noted. Second, the animal's recapture weight should always be compared with the original capture weight; any consistent weight loss between captures should be regarded as a possible indication that the radio package is hindering the animal's movements. Such problems are particularly important in radio-tracking studies of predation or mortality. In a study of deer mortality (Nelson and Mech 1981), a colleague and I found that collars taped with yellow to allow us to see the collar from the air also allowed hunters to see the deer more easily, and thus our early study was biased as far as hunter kills were concerned. Further details on the adverse effects of radio-tracking can be found in Amlaner (1978) and MacDonald and Amlaner (1980).

Physiological Telemetry

Most of this booklet has been devoted to a discussion of radio-tracking and the use of radio telemetry to help locate animals. However, it also is possible to use small radio transmitters to study such physiological parameters as deep-body temperature, heart rate, blood pressure, respiration, and peristalsis and to obtain electroencephalograms, electrocardiograms, and electromyograms (Van Citters and Franklin 1966; Van Citters et al. 1966; Amlaner 1978). Studies of these factors require the use of special equipment that is usually implanted in the animal and that has relatively limited range (Folk 1966). Therefore, most investigations of these parameters using telemetry are conducted with captive animals. Nevertheless, the technology is fast developing, and it is now practical to monitor for extended periods heart rate in some free-ranging ungulates (Cupal et al. 1976; MacArthur et al. 1979) and even in some invertebrates (Wolcott 1980a). Anyone interested should consult the literature on the subject, especially Mackay (1970), Ysenbrandt and others (1976), Amlaner (1978), Cochran (1980), and Amlaner and MacDonald (1980), and should write to the various commercial companies for catalogs of their equipment. (See also p. 13.)

Considerations in Starting a Radio-Tracking Program

Radio-tracking is a revolutionary technique and one that most wildlife biologists will find of considerable use. It is also different from the techniques with which most wildlife biologists have experience; therefore, a certain amount of practice is needed before one undertakes a radio-tracking project. Many beginners ask experienced radio-trackers to help them get started or visit radio-tracking projects in order to gain experience with them. A relatively small investment in one of these approaches pays great dividends in the resulting project. The following are several other considerations that should be made in setting up a radio-tracking project.

It is usually most efficient to buy the best-quality equipment. Capturing most study animals is difficult and requires considerable effort. If a radio collar fails prematurely or has insufficient range, much of this effort is wasted. When aircraft are involved in the radio-tracking, this point is particularly important, for every extra minute spent searching for a signal means considerable money is being wasted.

One must become expert in livetrapping the subject animals. Some species are easy to work with; others are very difficult, and special consultants may have to be hired or invited to help with the captures.

Since animals must be livetrapped to be radio-tagged, it makes sense to examine the study animals as intensively as possible, weighing them; taking various measurements; determining their age, sex, and condition; and collecting such physiological indicators of condition as blood samples, hair samples, urine samples, or any other specimens that might be available from any particular species (Mech 1980).

One must carefully design the radio-tracking study so as to match its objectives with the radio-tracking equipment available and the accuracy of that equipment. For example, in highly heterogeneous habitats, it may be very difficult to use radio-tracking to study habitat selection by animals, because radio-tracking is often too inaccurate to allow a determination of an animal's location precise enough to establish which type of habitat it might occupy.

One must make sure that it is possible to capture and to radio-track as many animals as might be needed in order to attain the study's objectives.

Best results can be obtained from radio-tracking by "living with" the radio-tagged animals, especially during the first few weeks after their release. Such intensity of data collection provides excellent insight into the lives of these animals and helps ensure the most effective and efficient use of the radio-tracking system.

COSTS

Radio-tracking can be relatively expensive in terms of initial cost. However, considering that it yields such a large amount of data and allows investigation of aspects of animal ecology and behavior that could not be studied without it, it is an excellent investment. Receivers generally cost from $700 to $1500 each, transmitters from $60 to $250, and antennas from $25 to $75. Unless a spare receiver is available for short periods from other projects, two are usually necessary for each project, one to be used should the other malfunction. Some money can be saved by leasing receivers for short-term studies and by purchasing unmounted transmitters and mounting them oneself in collars, harnesses, or other attachment devices. Nevertheless, in the long run, it is probably best to purchase the finished package unless the species to be studied is quite different from those for which commercial packages are readily available. When a researcher has more time and talent than funds, he or she can attempt to build the transmitters and save considerable money. However, this approach is usually less reliable and in the long run less satisfactory than purchasing the transmitters commercially (p. 49).

ORDERING EQUIPMENT

The names and addresses of various companies supplying radio-tracking equipment are listed in Appendix III. Most of them will send free catalogs upon request, and these should be studied before any equipment is ordered. It is usually necessary to order radio-tracking equipment several months before the study is scheduled to begin. If time allows, it is worthwhile to write the companies and to state very explicitly the following information: (1) the species to be studied, (2) the number of study animals, (3) the objectives of the study, (4) the geographic location of the study, (5) the type of topography and vegetation, (6) the annual variations in temperature, (7) the neck circumference or other pertinent anatomical measurements of the study animal, (8) the minimum and optimum transmitting range required, (9) the minimum and optimum transmitting life required, (10) the maximum and desirable radio-package weight (including attachment material), (11) the transportation methods to be used for obtaining signals (aircraft, foot, elephant, truck, jeep, etc.), (12) any local frequency regulations, and (13) any special foreseeable problems or potential compromises.

It is always wise to anticipate the need for spare parts so that the loss or damage of parts will not result in suspending data collection. For example, each project should have at least two receivers unless there is a nearby project from which a spare could be borrowed. Antennas and cables are other pieces of equipment that break easily and for which there should be spares.

LITERATURE CITED

Amlaner, C. J., Jr. 1978. Biotelemetry from free ranging animals. *In* B. Stonehouse (ed.). Animal Marking: Recognition Marking of Animals in Research, pp. 205-228. Macmillan, London.

Amlaner, C. J., Jr., and D. W. MacDonald (eds.). 1980. A Handbook on Biotelemetry and Radio Tracking. Pergamon Press, Oxford.

Anderka, F. W. 1980. Modulators for miniature tracking transmitters. *In* C. J. Amlaner, Jr., and D. W. MacDonald (eds.), A Handbook on Biotelemetry and Radio Tracking, pp. 181-184. Pergamon Press, Oxford.

Anderson, F., and P. O. De Moor. 1971. A system for radio-tracking monkeys in dense bush and forest. Journal of Wildlife Management 35: 636-643.

Ballard, W. B. 1982. Gray wolf-brown bear relationships in the Nelchina basin of south-central Alaska. *In* F. H. Harrington and P. C. Paquet (eds.), Wolves of the World, pp. 71-80. Noyes, Park Ridge, N.J.

Ballard, W. B., T. H. Spraker, and K. P. Taylor. 1981. Causes of neonatal moose calf mortality in south central Alaska. Journal of Wildlife Management 45:293-313.

Banks, E. M., R. J. Brooks, and J. Schnell. 1975. A radiotracking study of home range and activity of the brown lemming (*Lemmus trimucronatus*). Journal of Mammalogy 56:888-901.

Beach, M. H., and T. J. Storeton-West. 1982. Design consideration and performance checks on a telemetry tag system. *In* L. L. Cheesman and R. B. Mitson (eds.), Telemetric Studies of Vertebrates, pp. 31-45. Zoological Society of London Symposium 49.

Beale, D. M., and A. D. Smith. 1973. Mortality of pronghorn antelope fawns in western Utah. Journal of Wildlife Management 37:343-352.

Berg, W. E., and D. W. Kuehn. 1982. Ecology of wolves in north-central Minnesota. *In* F. H. Harrington and P. C. Paquet (eds.), Wolves of the World, pp. 4-11. Noyes, Park Ridge, N.J.

Bertram, B. 1980. The Serengeti radio-tracking program, 1971-1973. *In* C. J. Amlaner, Jr., and D. W. MacDonald (eds.), A Handbook on Biotelemetry and Radio Tracking, pp. 625-631. Pergamon Press, Oxford.

Buechner, H. K., F. C. Craighead, Jr., J. J. Craighead, and C. E. Cote. 1971.

Satellites for research on free roaming animals. BioScience 21:1201-1205.

Brand, C. J., R. H. Vowles, and L. B. Keith. 1975. Snowshoe hare mortality monitored by telemetry. Journal of Wildlife Management 39:741-747.

Bray, O. E., and G. W. Corner. 1972. A tail clip for attaching transmitters to birds. Journal of Wildlife Management 36:640-642.

Bray, O. E., R. E. Johnson, and A. L. Kolz. 1975. A removable car-top antenna system for radio-tracking birds. Bird-Banding 46:15-18.

Carbyn, L. N. 1982. Incidence of disease and its potential role in the population dynamics of wolves in Riding Mountain National Park, Manitoba. In F. H. Harrington and P. C. Paquet (eds.), Wolves of the World, pp. 106-116. Noyes, Park Ridge, N.J.

Carr, A. 1965. The navigation of the green turtle. Scientific American 212:79-86.

Charles-Dominique, P. 1977. Urine marking and territoriality in *Galago alleni* (Waterhouse, 1837—Lorisoidea, Primates): A field study by radio-telemetry. Zeitschrift für Tierpsychology 43:113-138.

Church, K. E. 1980. Expanded radio-tracking potential in wildlife investigations with the use of solar transmitters. In C. J. Amlaner, Jr., and D. W. MacDonald (eds.), A Handbook on Biotelemetry and Radio Tracking, pp. 247-250. Pergamon Press, Oxford.

Cochran, W. W. 1972. Long-distance tracking of birds. In S. R. Galler, K. Schmidt-Koenig, G. J. Jacobs, and R. E. Belleville (eds.), Animal Orientation and Navigation, pp. 39-59. NASA SP-262, U.S. Government Printing Office, Washington, D.C.

Cochran, W. W. 1975. Following a migrating peregrine from Wisconsin to Mexico. Hawk Chalk 14:28-37.

Cochran, W. W. 1980. Wildlife telemetry. In S. D. Schemnitz (ed.), Wildlife Management Techniques Manual (4th ed.), pp. 507-520. Wildlife Society, Washington, D.C.

Cochran, W. W., and R. D. Lord, Jr. 1963. A radio-tracking system for wild animals. Journal of Wildlife Management 27:9-24.

Cochran, W. W., G. G. Montgomery, and R. R. Graber. 1967. Migratory flights of *Hylocichla* thrushes in spring: A radiotelemetry study. Living Bird 6:213-225.

Cochran, W. W., D. W. Warner, J. R. Tester, and V. B. Kuechle. 1965. Automatic radio-tracking system for monitoring animal movements. BioScience 15:98-100.

Cook, R. S., M. White, D. O. Trainer, and W. C. Glazener. 1967. Radio-telemetry for fawn mortality studies. Wildlife Disease Association Bulletin 3:160-165.

Cook, R. S., M. White, D. O. Trainer, and W. C. Glazener. 1971. Mortality of young white-tailed deer fawns in south Texas. Journal of Wildlife Management 35:47-56.

Corner, G. W., and E. W. Pearson. 1972. A miniature 30-MHz collar transmitter for small animals. Journal of Wildlife Management 36:657-661.

Covich, A. 1977. Shapes of foraging areas used by radio-monitored crayfish. American Zoologist 17:205 (abstract).

Craighead, F. C., Jr., and J. J. Craighead. 1972. Grizzly bear prehibernation and denning activities as determined by radio-tracking. Wildlife Monographs 32.

Cupal, J. J., R. W. Weeks, and C. Kaltenbach. 1976. A heart rate and biotelemetry system for use on wild big game animals. *In* Proceedings of the Third International Symposium on Biotelemetry, pp. 219-222. Pacific Grove, Calif.

Deat, A., C. Mauget, R. Mauget, D. Maurel, and A. Sempere. 1980. The automatic, continuous and fixed radio-tracking system of the Chize Forest: Theoretical and practical analysis. *In* C. J. Amlaner, Jr., and D. W. MacDonald (eds.), A Handbook on Biotelemetry and Radio Tracking, pp. 439-451. Pergamon Press, Oxford.

Donaldson, N. de N. 1980. Encapsulation and packaging of implanted components. *In* C. J. Amlaner, Jr., and D. W. MacDonald (eds.), A Handbook on Biotelemetry and Radio Tracking, pp. 217-224. Pergamon Press, Oxford.

Dumke, R. T., and C. M. Pils. 1973. Mortality of radio-tagged pheasants on the Waterloo wildlife area. Wisconsin Department of Natural Resources Technical Bulletin 72.

Eliassen, E. 1960. A method for measuring the heart rate and stroke/pulse pressures of birds in normal flight. Arbok University of Bergen, Matenatisk Naturvitenskapelig 12:1-22.

Evans, W. E. 1971. Orientation behavior of delphinids: Radio-telemetry studies. Annals of the New York Academy of Sciences 188:142-160.

Floyd, T. J., L. D. Mech, and M. E. Nelson. 1979. An improved method of censusing deer in deciduous-coniferous forests. Journal of Wildlife Management 43:258-261.

Floyd, T. J., L. D. Mech, and M. E. Nelson. 1982. Deer in forested areas. *In* D. E. Davis (ed.), CRC Handbook of Census Methods for Terrestrial Vertebrates, pp. 254-256. CRC Press, Boca Raton, Fla.

Folk, G. E., Jr. 1966. Telemetered physiological measurements of bears in winter dens. Third Hibernation Symposium. Oliver and Boyd, Edinburgh.

Franzmann, A. W., C. C. Schwarz, and R. O. Peterson. 1980. Moose calf mortality in summer on the Kenai Peninsula, Alaska. Journal of Wildlife Management 44:764-768.

Fritts, S. H., and L. D. Mech. 1981. Dynamics, movements, and feeding ecology of a newly protected wolf population in northwestern Minnesota. Wildlife Monographs 80.

Fritts, S. H., W. J. Paul, and L. D. Mech. 1984. Movements of translocated wolves in Minnesota. Journal of Wildlife Management.

Fuller, T. K., and L. B. Keith. 1980. Wolf population dynamics and prey relationships in northeastern Alberta. Journal of Wildlife Management 44:583-602.

Gilmer, D. S., L. M. Cowardin, R. L. Duval, L. M. Mechlin, C. W. Shaiffer, and V. B. Kuechle. 1981. Procedures for the use of aircraft in wildlife biotelemetry studies. U.S. Fish and Wildlife Service Resource Publication 140.

Gilmer, D. S., V. B. Kuechle, and I. J. Ball, Jr. 1971. A device for monitoring radio-marked animals. Journal of Wildlife Management 35:829-832.

Hallberg, D. L., F. J. Janza, and G. R. Trapp. 1974. A vehicle-mounted directional antenna system for biotelemetry monitoring. California Fish and Game 60:172-177.

Harding, P. J. R., F. S. Chute, and A. C. Doell. 1976. Increasing battery reliability for radio transmitters. Journal of Wildlife Management 40:357-358.

Harrington, F. H., and L. D. Mech. 1982. Patterns of homesite attendance in two Minnesota wolf packs. In F. H. Harrington and P. C. Paquet (eds.), Wolves of the World, pp. 81-105. Noyes, Park Ridge, N.J.

Heezen, K. L., and J. R. Tester. 1967. Evaluation of radio-tracking by triangulation with special reference to deer movements. Journal of Wildlife Management 31:124-141.

Hoskinson, R. L. 1976. The effect of different pilots on aerial telemetry error. Journal of Wildlife Management 40:137-139.

Hoskinson, R. L., and L. D. Mech. 1976. White-tailed deer migration and its role in wolf predation. Journal of Wildlife Management 40:429-441.

Hutton, T. A., R. E. Hatfield, and C. C. Watt. 1976. A method for orienting a mobile radiotracking unit. Journal of Wildlife Management 40:192-193.

Ikeda, K., and M. Oshima. 1971. A wireless tracking system for behavioral studies of snakes. Snake 3:14-19.

Inglis, J. M. 1981. The forward-null twin-Yagi antenna array for aerial radiotracking. Wildlife Society Bulletin 9:222-225.

Jansen, D. K. 1982. A new potting material for radio-telemetry packages. Copeia 1982:189.

Kenward, R. E., G. J. M. Hirons, and F. Ziesemer. 1982. Devices for telemetering the behaviour of free-living birds. In C. L. Cheeseman and R. B. Mitson (eds.), Telemetric Studies of Vertebrates, pp. 129-137. Zoological Society of London Symposium 49.

Kephart, D. G. 1980. Inexpensive telemetry techniques for reptiles. Journal of Herpetology 14:285-290.

Kimmich, H. P. 1980. Artifact free measurement of biological parameters: Biotelemetry, a historical review and layout of modern developments. In C. J. Amlaner, Jr., and D. W. MacDonald (eds.), A Handbook of Biotelemetry and Radio Tracking, Pergamon Press, Oxford.

Kjos, C. G., and W. W. Cochran. 1970. Activity of migrant thrushes as determined by radio-telemetry. Wilson Bulletin 82:225-226.

Knowlton, F. F., P. E. Martin, and J. C. Haug. 1968. A telemetric monitor for determining animal activity. Journal of Wildlife Management 32:943-948.

Kohn, B. E., and J. J. Mooty. 1971. Summer habitat of white-tailed deer in north-central Minnesota. Journal of Wildlife Management 35:476-487.

Kolenosky, G. B. 1972. Wolf predation on wintering deer in east-central Ontario. Journal of Wildlife Management 36:357-369.

Kolz, A. L. 1975. Mortality-censusing wildlife transmitters. *In* Proceedings of the Twelfth International ISA Biomedical Sciences Instrumentation Symposium (April 28-30, 1975), pp. 57-60. Denver, Colo.

Kolz, A. L., G. W. Corner, and R. E. Johnson. 1973. A multiple-use wildlife transmitter. U.S. Fish Service Special Scientific Report on Wildlife 163.

Kolz, A. L., G. W. Corner, and H. P. Tietjen. 1972. A radio-frequency beacon transmitter for small mammals. Journal of Wildlife Management 36:177-179.

Kolz, A. L., and R. E. Johnson. 1975. An elevating mechanism for mobile receiving antennas. Journal of Wildlife Management 39:819-820.

Kolz, A. L., J. W. Lentfer, and H. G. Fallek. 1980. Satellite radio tracking of polar bears instrumented in Alaska. *In* C. J. Amlaner, Jr., and D. W. MacDonald (eds.), A Handbook on Biotelemetry and Radio Tracking, pp. 743-752. Pergamon Press, Oxford.

Kuechle, V. B. 1967. Batteries for biotelemetry and other applications. American Institute of Biological Sciences/Bio Instrumentation Advisory Council Information Module M 10.

Kuechle, V. B. 1982. State of the art of biotelemetry in North America. *In* C. L. Cheeseman and R. B. Mitson (eds.), Telemetric Studies of Vertebrates, pp. 1-18. Zoological Society of London Symposium 49.

Kuechle, V. B., D. P. DeMaster, and D. B. Siniff. 1979. State of the art and needs of the Earthplatform. Proceedings of the International Symposium on Remote Sensing Environment 13:505-518.

Legler, W. K. 1979. Telemetry. *In* M. Harless and H. Morlock (eds.), Turtles: Perspectives and Research, pp. 61-72. John Wiley and Sons, New York.

LeMunyan, D. C., W. White, E. Nybert, and J. J. Christian. 1959. Design of a miniature radio transmitter for use in animal studies. Journal of Wildlife Management 23:107-110.

Lord, R. D., F. C. Bellrose, and W. W. Cochran. 1962. Radio-telemetry of the respiration of a flying duck. Science 137:39-40.

Lotimer, J. S. 1980. A versatile coded wildlife transmitter. *In* C. J. Amlaner, Jr., and D. W. MacDonald (eds.), A Handbook on Biotelemetry and Radio Tracking, pp. 185-191. Pergamon Press, Oxford.

Lovett, J. W., and E. P. Hill. 1977. A transmitter syringe for recovery of immobilized deer. Journal of Wildlife Management 41:313-315.

MacArthur, R. A., R. H. Johnston, and V. Geist. 1979. Factors influencing heart rate in free-ranging bighorn sheep: A physiological approach to the studies of wildlife harassment. Canadian Journal of Zoology 57: 2010-2021.

MacDonald, D. W., and C. J. Amlaner, Jr. 1980. A practical guide to radio-tracking. *In* C. J. Amlaner, Jr., and D. W. MacDonald (eds.), A Handbook on Biotelemetry and Radio Tracking, pp. 143-159. Pergamon Press, Oxford.

Mackay, R. S. 1970. Bio-medical Telemetry (2nd ed.). John Wiley and Sons, New York.

Marshall, W. H., G. W. Gullion, and R. G. Schwab. 1962. Early summer activities of porcupines as determined by radio-positioning techniques. Journal of Wildlife Management 26:75-79.

Marshall, W. H., and J. J. Kupa. 1963. Development of radio-telemetry techniques for ruffed grouse studies. Transactions of the North American Wildlife Conference 28:443-456.

Mech, L. D. 1967. Telemetry as a technique in the study of predation. Journal of Wildlife Management 31:492-496.

Mech, L. D. 1974. Current techniques in the study of elusive wilderness carnivores. International Congress on Game Biology 11:315-322.

Mech, L. D. 1977a. Population trend and winter deer consumption in a Minnesota wolf pack. In R. L. Phillips and C. Jonkel (eds.), Proceedings of the 1975 Predator Symposium, pp. 55-83. Montana Forest and Conservation Experiment Station, Missoula.

Mech, L. D. 1977b. Productivity, mortality, and population trend in wolves from northeastern Minnesota. Journal of Mammalogy 58:559-574.

Mech, L. D. 1980. Making the most of radio-tracking: A summary of wolf studies in northeastern Minnesota. In C. J. Amlaner, Jr., and D. W. MacDonald (eds.), A Handbook on Biotelemetry and Radio Tracking, pp. 85-95. Pergamon Press, Oxford.

Mech, L. D., D. M. Barnes, and J. R. Tester. 1968. Seasonal weight changes, mortality, and population structure of raccoons in Minnesota. Journal of Mammalogy 49:63-73.

Mech, L. D., R. C. Chapman, W. W. Cochran, L. Simmons, and U. S. Seal. In press. A radio-triggered anesthetic-dart collar for recapturing free-ranging mammals. Bulletin of the Wildlife Society.

Mech, L. D., and L. D. Frenzel, Jr. 1971. Ecological studies of the timber wolf in northeastern Minnesota. U.S. Department of Agriculture Forest Service Research Paper NC-52. North Central Forest Experiment Station, St. Paul, Minn.

Mech, L. D., K. L. Heezen, and D. B. Siniff. 1966. Onset and cessation of activity in cottontail rabbits and snowshoe hares in relation to sunset and sunrise. Animal Behavior 14:410-413.

Mech, L. D., and M. Korb. 1978. An unusually long pursuit of a deer by a wolf. Journal of Mammalogy 59:860-861.

Mech, L. D., V. B. Kuechle, D. W. Warner, and J. R. Tester. 1965. A collar for attaching radio transmitters on rabbits, hares and raccoons. Journal of Wildlife Management 29:898-902.

Melquist, W. E., and M. G. Hornocker. 1979. Methods and techniques for studying and censusing river otter populations. University of Idaho Forest Wildlife and Range Experiment Station Technical Report 8.

Mineau, P., and D. Madison. 1977. Radio-tracking of Peromyscus leucopus. Canadian Journal of Zoology 55:465-468.

Nelson, M. E. 1979. Home range location of white-tailed deer. U.S. Department of Agriculture Forest Service Research Paper NC-173. North Central Forest Experiment Station, St. Paul, Minn.

Nelson, M. E., and L. D. Mech. 1981. Deer social organization and wolf predation in northeastern Minnesota. Wildlife Monographs 77:1-53.

Nicholls, T. H., and D. W. Warner. 1968. A harness for attaching radio transmitters to large owls. Bird-Banding 39:209-214.

Nicholls, T. H., and D. W. Warner. 1972. Barred owl habitat use as determined by radiotelemetry. Journal of Wildlife Management 36:213-224.

Osgood, D. W. 1970. Thermoregulation in water snakes studied by telemetry. Copeia 1970:568-571.

Patton, D. R., D. W. Beatty, and R. H. Smith. 1973. Solar panels: An energy source for radio transmitters on wildlife. Journal of Wildlife Management 37:236-238.

Pienkowski, E. C. 1965. Predicting transmitter range and life. BioScience 15:115-118.

Raim, A. 1978. A radio transmitter attachment for small passerine birds. Bird-Banding 49:326-332.

Rawson, K. S., and P. H. Hartline. 1964. Telemetry of homing behavior by the deer-mouse, *Peromyscus*. Science 146:1596-1597.

Rogers, L. L., and L. D. Mech. 1981. Interactions of wolves and black bears in northeastern Minnesota. Journal of Mammalogy 62:434-436.

Sargeant, A. B. 1980. Approaches, field considerations and problems associated with radio-tracking carnivores. *In* C. J. Amlaner, Jr., and D. W. MacDonald (eds.), A Handbook on Biotelemetry and Radio Tracking, pp. 57-63. Pergamon Press, Oxford.

Schubauer, J. P. 1981. A reliable radio-telemetry tracking system suitable for studies of Chelonians. Journal of Herpetology 15:117-120.

Servheen, C., T. T. Thier, C. J. Jonkel, and D. Beatty. 1981. An ear-mounted transmitter for bears. Wildlife Society Bulletin 9:56-57.

Siniff, D. B., R. Reichle, R. Hofman, and D. Kuehn. 1975. Movements of Weddell seals in McMurdo Sound, Antarctica, as monitored by telemetry. Rapport International Counsel for the Exploration of the Sea 169:387-393.

Skiffins, R. M. 1982. Regulatory control of telemetric devices used in animal studies. *In* C. L. Cheeseman and R. B. Mitson (eds.), Telemetric Studies of Vertebrates, pp. 19-30. Zoological Society of London Symposium 49.

Slater, L. E. 1963. Biotelemetry: The Use of Telemetry in Animal Behavior and Physiology in Relation to Ecological Problems. Proceedings of the Interdisciplinary Conference. Pergamon Press, Oxford.

Smith, E. N. 1975. Thermoregulation of the american alligator (*Alligator mississippiensis*). Physiological Zoology 48:177-194.

Smith, H. R. 1980. Growth, reproduction and survival in *Peromyscus leucopus* carrying intraperitoneally implanted transmitters. *In* C. J. Amlaner, Jr., and D. W. MacDonald (eds.), A Handbook on Biotelemetry and Radio-Tracking, pp. 367-374. Pergamon Press, Oxford.

Smith, H. R., and G. Whitney. 1977. Intraperitoneal transmitter implants:

Their biological feasibility for studying small mammals. *In* Proceedings, First International Conference on Wildlife Biotelemetry, pp. 109-117. Laramie, Wyo.

Smith, G. J., J. R. Cary, and O. J. Rongstad. 1981. Sampling strategies for radio-tracking coyotes. Wildlife Society Bulletin 9:88-93.

Stasko, A. B. 1975. Underwater biotelemetry, annotated bibliography. Canada Fisheries and Marine Service Technical Report 534.

Stasko, A. B., and D. G. Pincock. 1977. Review of underwater biotelemetry, with emphasis on ultrasonic techniques. Journal of the Fisheries Research Board of Canada 34:1262-1285.

Stoddart, L. C. 1970. A telemetric method for detecting jackrabbit mortality. Journal of Wildlife Management 34:501-507.

Stoneburner, D. L. 1982. Satellite telemetry of loggerhead sea turtle movement in the Georgia Bight. Copeia 1982:400-408.

Tester, J. R., D. W. Warner, and W. W. Cochran. 1964. A radio-tracking system for studying movements of deer. Journal of Wildlife Management 28:42-45.

Thomas, D. W. 1980. Plans for a lightweight inexpensive radio transmitter. *In* C. J. Amlaner, Jr., and D. W. MacDonald (eds.), A Handbook on Biotelemetry and Radio Tracking, pp. 175-179. Pergamon Press, Oxford.

Trent, T. T., and O. J. Rongstad. 1974. Home range and survival of cottontail rabbits in southwestern Wisconsin. Journal of Wildlife Management 38:459-472.

Turkowski, F. J., and L. D. Mech. 1968. Radio-tracking a young male raccoon. Journal of the Minnesota Academy of Science 35(1):33-38.

Van Camp, J., and R. Gluckie. 1979. A record long-distance move by a wolf (*Canis lupus*). Journal of Mammalogy 60:236-237.

Van Citters, R. L., and D. L. Franklin. 1966. Telemetry of blood pressure in free-ranging animals via an intravascular gauge. Journal of Applied Physiology 21:1633-1636.

Van Citters, R. L., and D. Franklin. 1969. Radio telemetry techniques for study of cardiovascular dynamics in ambulatory primates. Annals of the New York Academy of Sciences 162:137-155.

Van Citters, R. L., W. S. Kemper, and D. L. Franklin. 1966. Blood pressure responses of wild giraffes studied by radio telemetry. Science 152:384-386.

Van Citters, R. L., O. A. Smith, D. L. Franklin, W. S. Kemper, and N. W. Watson. 1967. Radio telemetry of blood flow and blood pressure in feral baboons: A preliminary report. *In* H. Vagtborg (ed.), The Baboon in Medical Research II, pp. 473-492. Proceedings of the Second International Symposium on the Baboon and Its Use as an Experimental Animal. University of Texas Press, Austin.

Verts, B. J. 1963. Equipment and techniques for radio-tracking striped skunks. Journal of Wildlife Management 27:325-339.

Walcott, C., J. L. Gould, and J. L. Kirschvink. 1979. Pigeons have magnets. Science 205:1027-1029.

Ward, A. L., and J. J. Cupal. 1979. Telemetered heart rates of three elk as affected by activity and human disturbance. *In* K. Downing (ed.), Dispersed Recreation and Natural Resources Management, pp. 47-56. Utah State University Logan.

Warner, D. W. 1967. Space tracks. *In* I. W. Knoblack (ed.), Readings in Biological Science (2nd ed.), pp. 221-228. Appleton-Century-Crofts, New York.

Weise, T. F., W. L. Robinson, R. A. Hook, and L. D. Mech. 1975. An experimental translocation of the eastern timber wolf. Audubon Conservation Report 5. U.S. Fish and Wildlife Service, Twin Cities, Minn.

White, G. C. 1979. Computer generated movies to display biotelemetry data. *In* Proceedings, Second International Conference on Wildlife Biotelemetry, pp. 210-214. Laramie, Wyo.

Whitehouse, S., and D. Steven. 1977. A technique for aerial radio tracking. Journal of Wildlife Management 41:771-775.

Williams, T. C., and E. J. Burke. 1973. Solar power for wildlife telemetry transmitters. American Birds 27:719-720.

Winter, J. D., V. B. Kuechle, D. B. Siniff, and J. R. Tester. 1978. Equipment and methods for radio-tracking freshwater fish. Agricultural Experiment Station, University of Minnesota Miscellaneous Report 152.

Wolcott, T. G. 1980a. Heart rate telemetry using micropower integrated circuits. *In* C. J. Amlaner, Jr., and D. W. MacDonald (eds.), A Handbook on Biotelemetry and Radio Tracking, pp. 279-286. Pergamon Press, Oxford.

Wolcott, T. G. 1980b. Optical and radio optical techniques for tracking nocturnal animals. *In* C. J. Amlaner, Jr., and D. W. MacDonald (eds.), A Handbook on Biotelemetry and Radio Tracking, pp. 333-338. Pergamon Press, Oxford.

Ysenbrandt, H. J. B., T. A. J. Selten, J. J. M. Verschuren, T. Kock, and H. P. Kimmich. 1976. Biotelemetry, literature survey of the past decade. Biotelemetry 3:145-250.

Zimmermann, F., H. Geraud, and P. Charles-Dominique. 1975. Le radio-tracking des vertebrates: Conseils et techniques d'utilisation. LaTerre et la Vie, 30:309-346.

Appendixes

APPENDIX I
General References

Amlaner, C. J., Jr. 1978. Biotelemetry from free ranging animals. *In* B. Stonehouse (ed.), Animal Marking: Recognition Marking of Animals in Research, pp. 205-228. Macmillan, London.

Amlaner, C. J., Jr., and D. W. MacDonald (eds.). 1980. A Handbook on Biotelemetry and Radio Tracking. Pergamon Press, Oxford.

Cochran, W. W. 1980. Wildlife telemetry. *In* S. D. Schemnitz (ed.), Wildlife Management Techniques Manual (4th ed.), pp. 507-520. Wildlife Society, Washington, D.C.

Klewe, H. J., and H. P. Kimmich (eds.). 1978. Biotelemetry IV. Proceedings of the Fourth International Symposium on Biotelemetry. Braunschweig, Germany.

MacDonald, D. W. 1978. Radio-tracking: Some applications and limitations. *In* B. Stonehouse (ed.), Animal Marking: Recognition Marking of Animals in Research, pp. 192-204. Macmillan, London.

Mackay, R. S. 1970. Bio-medical Telemetry (2nd ed.). John Wiley and Sons, New York.

Offner, F. F. 1967. Electronics for Biologists. McGraw-Hill, New York.

Slater, L. E. 1963. Biotelemetry: The Use of Telemetry in Animal Behavior and Physiology in Relation to Ecological Problems. Proceedings of the Interdisciplinary Conference. Pergamon Press, Oxford.

Stasko, A. B., and D. G. Pincock. 1977. Review of underwater biotelemetry, with emphasis on ultrasonic techniques. Journal of the Fisheries Research Board of Canada 34:1262-1285.

Will, G. B., and E. F. Patric. 1972. A contribution toward a bibliography on wildlife telemetry and radio tracking. New York Department of Environmental Conservation, Delmar Wildlife Research Laboratory, Delmar, N.Y.

Ysenbrandt, H. J. B., T. A. J. Selten, J. J. M. Verschuren, T. Kock, and H. P. Kimmich. 1976. Biotelemetry, literature survey of the past decade. Biotelemetry 3:145-250.

Zimmermann, F., H. Geraud, and P. Charles-Dominique. 1975. Le radio-tracking des vertebrates: Conseils et techniques d'utilisation. LaTerre et la Vie, 30:309-346.

APPENDIX II
Exercises to Practice
Radio-Tracking

1. Place a transmitter in a conspicuous location that can be seen from a distance; move about 100 m away, tune in the transmitter on your receiver, and slowly wave your antenna around 360° to determine how the signal changes as the antenna points toward the transmitter. Repeat for several gain or volume settings.
2. Have another person hide a transmitter about 0.5 to 1.0 km away, and then try to find it by homing.
3. Have another person hide a transmitter along a road about 2 to 10 km away from camp but about 0.5 km from the road (the person should remember exactly where it was hidden). The person tells you the direction from camp and which road the transmitter is near. As you drive along the road, stop every kilometer to listen for the signal. Try to home in on and find the transmitter.
4. Try this exercise with three people. In a large open field, have person number 1 hide a transmitter about 0.5 km away from the starting point. Person number 2 tries to take a bearing on the transmitter from a distance and sends person number 1 to walk along that bearing. Person number 1 should be sent toward a landmark, for example, a prominent tree in the background, toward which the bearing points. Person number 2 then moves about 300 m, takes another bearing, and sends person number 3 to walk along it. Person number 1 and person number 3 should coordinate their travel so that they start about the same time. They should stop at the point where their paths cross. That point will be the radio-tracker's estimate of where the transmitter is hidden. A rock or some other mark should be placed there. The two people should return to the radio-tracker. This procedure should be repeated at least twice more, with a marker placed at each crossing point. Then, the radio-tracker should home in on the transmitter and determine how close each estimated point was to the transmitter. This information indicates the accuracy of the antennas and the skill of the radio-tracker. The radio-tracker should repeat this test as many times as necessary until his or her skill is as good as possible.

APPENDIX III
Suppliers of
Radio-Tracking Equipment*

UNITED STATES

Advanced Telemetry Systems, Inc.
23859 Northeast Highway 65
Bethel, MN 55005
(612) 434-5040

AVM Instrument Company
6575 Trinity Court
Dublin, CA 94566
(415) 829-5030

Cedar Creek Bioelectronics Lab
University of Minnesota
Bethel, MN 55005
(612) 434-7361

CompuCap
P.O. Box 864
Minneapolis, MN 55440
(612) 642-1368

Custom Electronics
2009 Silver Court West
Urbana, IL 61801
(217) 344-3460

Custom Telemetry and Consulting
185 Longview Drive
Athens, GA 30605
(404) 548-1024

Dav-Tron
400 Penn Avenue South
Minneapolis, MN 55405
(612) 377-5244, (612) 377-7770

L. L. Electronics
P.O. Box 247
Mahomet, IL 61853
(217) 586-2132

Ocean Applied Research
  Corporation
10447 Roselle Street
San Diego, CA 92121
(619) 453-4013

Smith-Root, Inc.
14014 Northeast Salmon Creek
  Avenue
Vancouver, WA 98665
(206) 573-0202

Stuart Enterprises
P.O. Box 310
124 Cornish Court
Grass Valley, CA 95945
(916) 273-9188

Telemetry Systems, Inc.
P.O. Box 187
Mequon, WI 53092
(414) 241-8335

Telonics
932 East Impala Avenue
Mesa, AZ 85204-6699
(602) 892-4444

Wildlife Materials, Inc.
R.R. 1
Giant City Road
Carbondale, IL 62901
(618) 549-6330

Wyoming Biotelemetry, Inc.
1225 Florida Avenue
Longmont, CO 80501
(303) 772-5948

*No endorsements of commercial suppliers should be inferred from this list.

EUROPE

Mariner Radar
Bridleway,
Campsheath,
Lowestoft
Suffolk, NR32 5DN
UNITED KINGDOM
(0502) 67195

Biotrack
c/o Monk's Wood Experimental
  Station
Abbott's Ripton
Huntingdon PE 172LS
UNITED KINGDOM

Index

INDEX

Abdomen, transmitter implanted in, 10, 44, 80

Accuracy, 32, 34, 43, 53, 58, 60, 75-79, 82, 96. *See also* Precision

Acrylic, 9, 42, 46

Adapter: for antenna connector, 25, 26, 31; female, 31; male, 31

Aerial census, aided by radio-tracking, 10

Aerial homing, 24, 61-71, 78

Aerial observation of radio-collared deer, 10, 79

Aerial radio-tracking, 24, 47, 54-57, 61-72, 78

Age of dispersal, studying, 5

Aircraft, 14, 15, 24, 25, 38, 54, 55, 57, 61-71, 78, 79, 81, 83

Alligators, 11

Altitude, 5, 11, 62, 64, 65, 66

Ambient noise and earphones, 24

Anatomical measurements, informing manufacturers about, 83

Animals: habits of, 3; locating, 3, 4-6, 8, 21, 51, 60, 61, 68-69; movements of, 3, 4-6, 12-13, 79; activity of, 6-8, 12, 26, 30-32, 51, 72-74; mortality, 8, 12, 51, 60, 79; capture of, 8, 14-15, 46, 47, 79, 81, 82; condition, measurement of, 8, 81; observation of, 9, 10, 60, 68-69, 78; anesthetizing, 10, 14, 15, 60, 79; lost, 24, 47, 56, 61, 64; absence/presence, monitoring by signal, 26, 30, 51, 72-74;

simulated, 37, 39, 41; disturbance, 51, 79; bed, confirming accuracy of, 78; captive, physiology of, 80; aging, need for, 81; examining study animals, 81; sexing, 81. *See also* Behavior

Antenna: directional, 3, 26, 27, 28, 39, 50, 51, 54, 56, 58, 61, 75; transmitting, 6, 15, 18, 20-21, 35, 36, 37, 39, 46; whip, 6, 20, 21, 36; receiving, 15, 21, 22, 25-32 *passim*, 38, 39, 40, 50-62 *passim*, 71-77, 82, 83; ideal, 20; length of, 20, 28, 36, 37; quarter-wave, 20; loop, 20-21, 36, 38; PL-259 connector, 25, 26; jacks, 25, 57; half-wave-length, 25-26; dipole, 26, 54; omnidirectional, 26; bidirectional, 26-28; boom, 27-28; yagi, 27, 28, 35, 38, 39, 40, 51-56 *passim*; Adcock, 28; collapsible, 28; elements of, 28, 36, 39, 40; pickup, 28, 52-53, 54; sensitivity of, 28; accessories, 28-29; tower, 29, 39, 51; mast, 29, 40, 51, 52, 53; outdoor cable, 29-30; compatability of connector, 29-31; optimum transmitting, 37; H-, 38, 54; rotatable, 39, 40, 41, 53, 60-61; hand-held, 39, 40, 50, 51, 58; mounting of, 39, 42, 54, 82; stacking of, 39, 51, 52-53; Cush-craft, 40; portable, 40; switch,

L. David Mech is a wildlife research biologist with the U.S. Fish and Wildlife Service and an adjunct professor in the department of ecology and behavioral biology and the department of entomology, fisheries, and wildlife at the University of Minnesota. Educated at Cornell and at Purdue University, where he earned a doctorate in ecology, Mech participated in the developmental work in radio-tracking at Minnesota in the early 1960s. He is the author of *The Wolves of Isle Royale* and *The Wolf: The Ecology and Behavior of an Endangered Species* (available in paperback from Minnesota).